Introduction to

VETERINARY ENTOMOLOGY

D. E. Bay
Texas A&M University

R. L. Harris
USDA/ARS (Retired)

ROBERT HARRIS PUBLISHER

Robert Harris Publisher

ISBN 0-9624083-0-1

Printed in the United States of America

SECOND EDITION

SECOND PRINTING

PREFACE

Introduction to Veterinary Entomology is intended to acquaint students and livestock producers with the most important species of insects, ticks and mites attacking livestock and poultry in the United States. The text includes general information on the biology, ecology and control of the pest as well as the nature of damage to the host. Specific chemical control recommendations are not included since these are reviewed and updated annually current information should be obtained from state extension personnel or other local specialists.

The authors gratefully acknowledge the United States Department of Agriculture (USDA), United States Department of Health, Education and Welfare, Center for Disease Control (CDC), University of California Cooperative Extension (CCEX), Pennsylvania State University Cooperative Extension Service (PCEX) and the Texas Agricultural Extension Service (TAEX) for permitting us to use the following publications in the preparation of this text.

Anonymous. 1960. Introduction to Arthropods of Public Health Importance. HEW Publication No. (CDC) 79-8139. 48 p.

Anonymous. 1974. Insecticides for the Control of Insects of Public Health Importance. HEW Publication No. (CDC) 79-8229. 71p.

Anonymous. 1978. Ticks of Public Health Importance and their Control. HEW Publication No. (CDC) 78-8142. 48 p.

Collison, C. H. 1978. Controlling Insects and Mites on Dairy and Beef Cattle. Pennsylvania State University Extension Service Special Circular 230. 26 p.

Collison, C. H. 1978. Controlling Insects and Mites on Sheep and Goats. Pennsylvania State University Extension Service Special Circular 244. 11 p.

Hoelscher, C. E. 1985. Extemal Parasites of Cattle. Texas Agricultural Extension Service B-1080. 24 p.

Hoelscher, C. E. 1985. External Parasites of Poultry. Texas Agricultural Extension Service B-1088. 14 p.

Loomis, E. C. 1978. The Common Parasites of Horses. University of California, Division of Agricultural Sciences Sale Publication 4006. 35 p.

Loomis, E. C. 1982. Common External Parasites and Pests of Livestock and Poultry in California. University of California Cooperative Extension Service Booklet 71 p.

Price, M. A., W. H. Newton, and P. J. Hamman. 1972. External Parasites of Texas Sheep and Goats. Texas Agricultural Extension Service MP-834. 11 p.

Price, M. A., W. H. Newton, and P. J. Hamman. 1967. Insect, Mite and Tick Parasites of Texas Horses. Texas Agricultural Extension Service MP-83. 8 p.

Suickland, R. K., R. R. Gerrish, J. L. Hourrigan, and G. O. Schubert. 1976. Ticks of Veterinary Importance. USDA Agricultural Handbook No. 485. 122 p.

We wish to express our appreciation to the following individuals for their reviews and comments: W. F. Chamberlain, J. A. DeVaney, W. A. Donahue, G. T. Fincher, J. E. George, O. H. Graham, F. S. Guillot, S. E. Kunz, J. A. Miller, M. A. Price, P. D. Teel and H. W. Van Cleave. We are especially indebted to Elizabeth Andrus who typed and retyped the numerous drafts.

Darrell E. Bay
Robert L. Harris

CONTENTS

CHAPTER 1

INSECTS AND OTHER ARTHROPODS

CLASSIFICATION

The lice, fleas, flies, ticks and mites affecting livestock, poultry and companion animals belong to the group of animals termed the Arthropoda. This group, or PHYLUM (plural: PHYLA), is characterized by segmented appendages, a segmented body and an exoskeleton of chitin. Arthropods represent an extremely diverse group; they occupy a wide variety of ecological habitats, exhibit extensive morphological differences and consume a broad assortment of foods. Approximately 80% of all described species of animals belong to this group. The phylum Arthropoda is further divided into CLASSES including the Crustacea (crabs, shrimps, copepods and others), Chilopoda (centipedes), Diplopoda (millipedes), Arachnida (spiders, scorpions, ticks and mites) and Insecta (insects). Figure 1.1 illustrates the predominance of insects, amounting to over 750,000 species, in the world fauna. By constrast, it is estimated that there are only about 33,000 living species of Chordata (fishes, amphibians, reptiles, birds and mammals).

The class Insecta, or Hexapoda, contains a number of major groupings termed ORDERS, such as the order Diptera, or true flies. Each order is made up of FAMILIES, such as the family Culicidae, or mosquitoes. Each family is comprised of one or more GENERA (singular: GENUS), such as the genus *Culex*; and each genus has one or more SPECIES, such as *pipiens*, the house mosquito, which may have several SUBSPECIES, such as *Culex pipiens pipiens*, the northern house mosquito, and *Culex pipiens quinquefasciatus*, the southern house mosquito. Thus, the scientific name usually consists of either two or three terms, the genus and species names, as in *Musca domestica* for the house fly, or the genus, species and subspecies names, as in *Culex pipiens quinquefasciatus*. Scientific names are always written in italics or are underlined; only the first, or generic name, begins with a capital letter.

This system of binomial nomenclature was consistently applied to animals by the Swedish biologist, Linnaeus, in the tenth edition of his classical book "*Systema Naturae*" in 1758. Linnaeus wrote in Latin, the scholarly language of his time, and the Latin names of his major groupings have been universally adopted, understood and used throughout the world. Thus, *Aedes aegypti* refers to the same species of mosquito to French, German or Spanish workers, regardless of their ability to read "moustique", "stechmucken" or "zancudo" in their respective languages. Frequently, the scientific name also includes the last name of the scientist, or AUTHORITY, who first described a particular species. The author's name in parentheses indicates that the species was originally described in another genus. For example, the yellow fever mosquito, *Aedes aegypti* (Linnaeus), was originaily described as *Culex aegypti* by Linnaeus in 1762, some 50 years before the present genus *Aedes* was first separated from the genus *Culex* by Meigen in 1818.

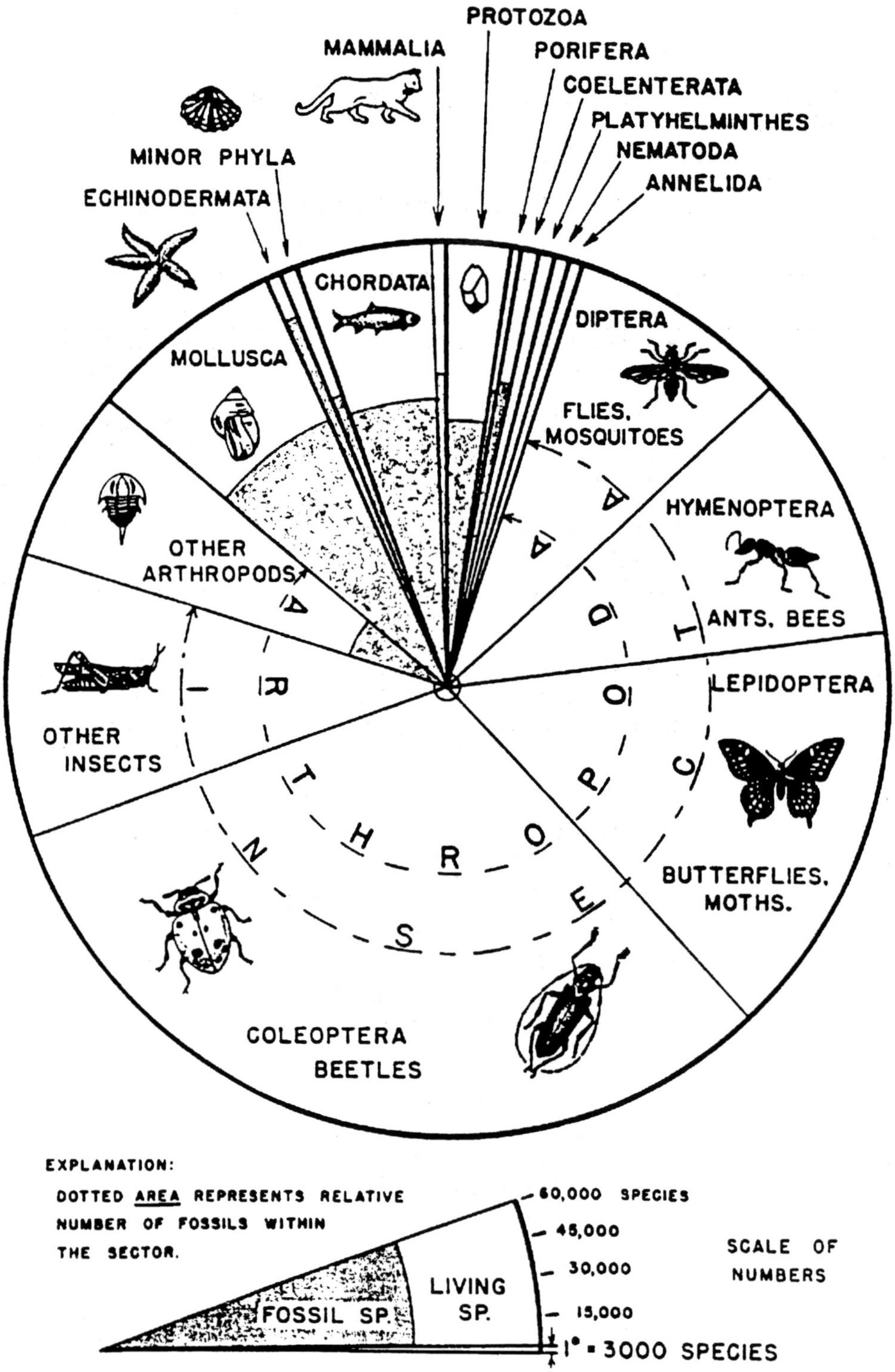

Figure 1.1 Relative numbers of known species, living and fossil, of various animal phyla. (From Muller and Campbell, 1954. Syst. Zool. 3:169-170).

A recapitulation of the steps involved in classifying a mosquito is demonstrated in Table 1 along with the comparable scientific classification of man.

TABLE 1 - THE CLASSIFICATION OF A MOSQUITO AND MAN

Name of Catagory	Mosquito	Man
KINGDOM	ANIMAL - Capable of motion, no chlorophyll.	ANIMAL - Capable of motion, no chlorophyll.
PHYLUM	ARTHROPODA - Jointed appendages, exoskeleton, dorsal heart, ventral nerve cord, cold blooded.	CHORDATA - Jointed appendages, endoskeleton, ventral heart, dorsal nerve cord.
CLASS	INSECTA - Three pairs of legs, one pair of antennae, wings, usually present.	MAMMALIA - Mammary glands for suckling young, hair, four-chambered heart, warm blooded.
ORDER	DIPTERA - Two wings, second pair of wings modified into halteres or balancers, sucking mouthparts, complete metamorphosis.	PRIMATES - Limbs elongate, "hands" and "feet" enlarged, often with a thumb, each of five digits with flattened or cupped nails.
FAMILY (Note family names end in "idae" in zoology)	Culicidae (mosquitoes)- Adults with scales on wings, elongate proboscis, wings usually longer than abdomen.	Hominidae - The family of man.
GENUS	*Culex* (Latin for mosquito) (Note that the generic name of any animal is capitalized and written in italics).	*Homo* (Latin for man).
SPECIES	*pipiens* (note that the species name is always written in small letters and in italics). This species name refers to the piping or whining sound of the flying mosquito.	*sapiens* (Latin for learned).
SUBSPECIES	*quinquefasciatus*	*sapiens*
AUTHORITY & DATE	Say, 1823	Linnaeus, 1758

THE CLASSES OF ARTHROPODS

The Class Chilopoda: Centipedes

The centipedes (Fig. 1.2) are commonly referred to as "hundred-leggers" although most species possess less than this number. Centipedes are characterized by one pair of antennae and two distinct body regions, the head and trunk. The trunk is composed of a group of similar segments, each of which bears a single pair of legs Certain species of centipedes are capable of injecting a poisonous venom, comparable to that of a bee or wasp, through a pair of poison claws arising from the first trunk segment immediately behind the head.

Figure 1.2. Class Chilopoda, a centipede. (CDC)

The Class Diplopoda: Millipedes

The millipedes (Fig. 1.3) also possess one pair of antennae and two body regions; however, they differ from centipedes in that each apparent trunk segment bears two pairs of legs. Millipedes are usually not injurious to man or animals, but they may cause damage to plant crops. A few species are capable of exuding or squirting a blistering fluid and several serve as intermediate hosts for certain parasitic worms.

Figure 1.3. Class Diplopoda, a millipede. (CDC)

The Class Crustacea: Crabs, Shrimps, Lobsters, Copepods and Sowbugs

Crustaceans (Fig. 1.4) differ from insects in that they have five or more pairs of legs and two pairs of antennae. In lobsters and related forms, the head and thorax are combined to form a cephalothorax; the remaining area is termed the abdomen. Crustaceans are represented by crabs, shrimps, barnacles, sowbugs and many other diverse forms. Members of this class constitute an important source of food to both man and many aquatic animals, and may also serve as intermedate hosts of certain species of parasitic worms.

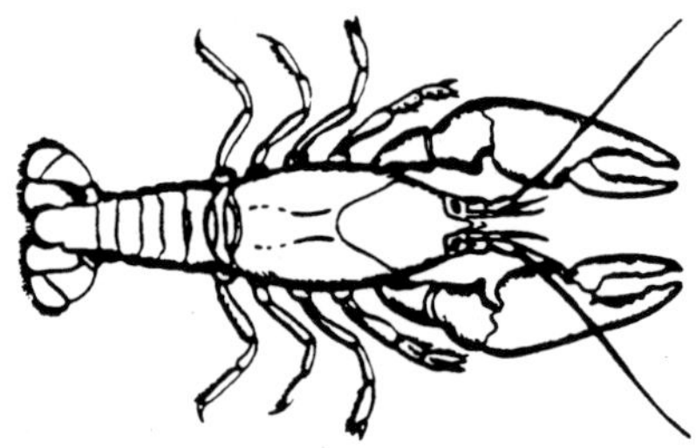

Figure 1.4. Class Crustacea, a crayfish. (CDC)

The Class Arachnida: Scorpions, Spiders, Ticks and Mites

Members of the class Arachnida may be easily distinguished by the fact that they lack antennae and possess four pairs of legs in the adult stage. The first two body regions (head and thorax) are fused to form a cephalothorax in some species while others, such as the ticks and mites, have these fused into a single body region. Respiration is accomplished by means of book-lungs or tubular tracheae. Although there are a number of orders in the class Arachnida, only three are of major importance to medical/veterinary entomology; these are the Scorpionida (scorpions), Araneae (spiders) and Acari (ticks and mites).

Order Scorpionida: Scorpions

Scorpions (Fig. 1.5) are arachnids with a segmented abdomen broadly joined to the cephalothorax. The long, slender tip of the abdomen bears a stinger, enclosing a venomous gland, used to paralyze prey as well as for defensive purposes. The large pincer-like palps, used to catch and hold prey, are carried in front of the first pair of legs. Scorpions frequently enter buildings and hide in cracks and similar dark, protected places. A number of species sting severely, and the venom of some tropical species may result in death.

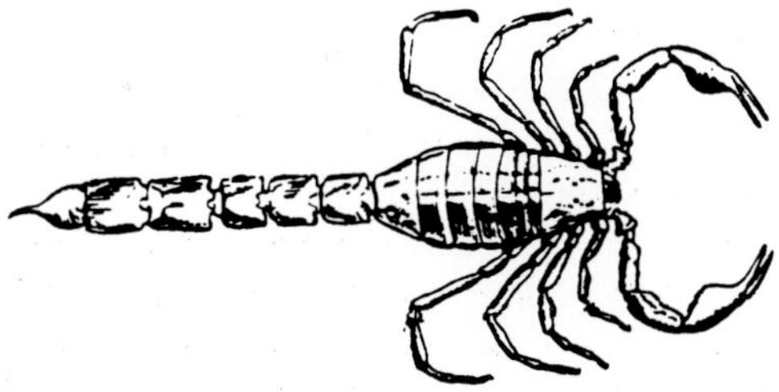

Figure 1.5. Order Scorpionida, a scorpion. (CDC)

Order Araneae: Spiders

Spiders differ from other arachnids in that the cephalothorax is joined to the abdomen by a slender waist, known technically as a pedicel, and the abdominal segmentation is either indistinct or absent. Most species are beneficial in that they are predaceous on many harmful insects. However some species of widow spiders, characterized by an hourglass configuration on the underside of the abdomen (genus *Latrodectus*), have a very venomous bite that may result in death. The black widow spider, *Lactrodectus mactans* (Fig. 1.6), occurs throughout the entire continental United States; the venom is a neurotoxin which affects the central nervous system. The venom of the brown recluse spider, *Loxosceles recluso* (Fig. 1.7), is of the haemolytic type; bites are usually localized resulting in mild to, severe localized necrosis or gangrene.

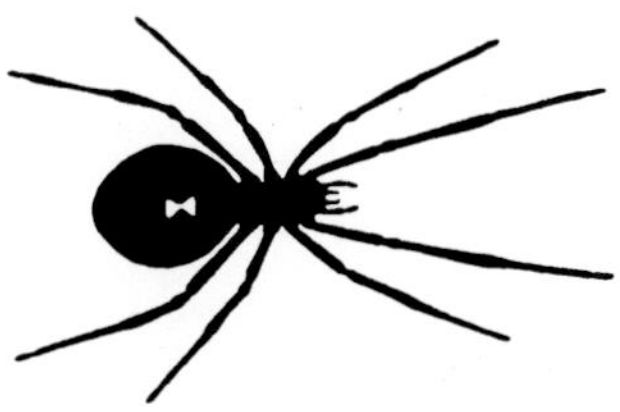

Figure 1.6. Order Araneae, black widow spider.

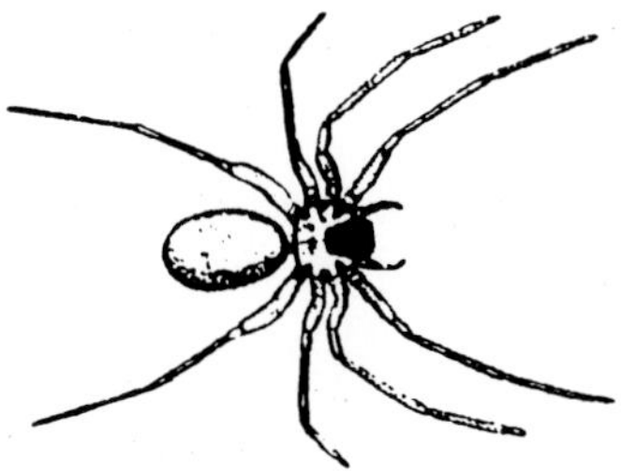

Figure 1.7. Order Araneae, brown recluse spider. (CDC)

Order Acari: Ticks and Mites

Ticks and mites (Fig. 1.8) have the head, thorax and abdomen combined into a single body region which is sac-like in form. Ticks possess a hypostome, provided with recurved teeth, which is used as a holdfast organ after the mouthparts, termed chelicerae, have made an incision in the host in order to feed. Mites are usually much smaller than ticks, and the hypostome, if present, is not armed with teeth. Ticks and mites have three pairs of legs in the first or larval stage, but nymphs and adults characteristically possess four pairs of legs. Next to the insects, ticks and mites constitute the most important group of arthropods affecting man and animals.

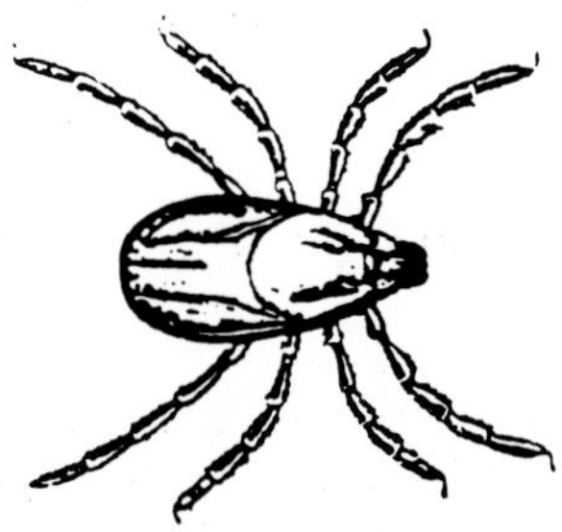

Figure 1.8. Order Acari, brown dog tick. (USDA)

The Class Insecta: Insects

Insects constitute the most important group of the phylum Arthropoda. Typical adult insects have one pair of antennae, three pairs of legs and three distinct body regions (head, thorax and abdomen). Many, but not all, adult insects possess wings; none of the other arthropod classes have wings. Some insects are vicious, blood-sucking pests while others are mechanical or biological carriers of many important diseases of man and animals. The class Insecta is divided into major groups, or orders, by different authorities. Members of each order have certain features in common, primarily the type of mouthparts, the number of pairs and type of wings, and the kind of metamorphosis, that distinguish them from all other species. Four of these orders are of major importance to veterinary entomology: Mallophaga (biting lice), Anoplura (sucking lice), Siphonaptera (fleas) and Diptera (flies). A few primitive insects, such as the silverfish, develop without metamorphosis. Others, such as cockroaches, lice and true bugs, develop by gradual or incomplete metamorphosis with three stages (egg, nymyh and adult) in their life histories. An even larger number of insects (fleas, flies, beetles, butterfiies and moths) develop by complete metamorphosis with four stages (egg, larva, pupa and adult) in their life cycles. These types of development are described in more detail in Chapter 3, "The Development of Insects". Following is a brief account of the orders of insects.

THE ORDERS OF INSECTS

INSECTS WITHOUT METAMORPHOSIS

Order Thysanura: Silverfish and Firebrats

Silverfish Fig. 1.9) and firebrats are characterized by chewing type mouthparts, scales on the body and three long appendages at the posterior end of the abdomen. Young possess the obvious structures of the adults, differing chiefly in size, color and sexual maturity. These insects are frequently found in dwellings where they may be destructive of property by feeding on clothing and starchy materials such as books or wallpaper, or on foodstuffs such as flour.

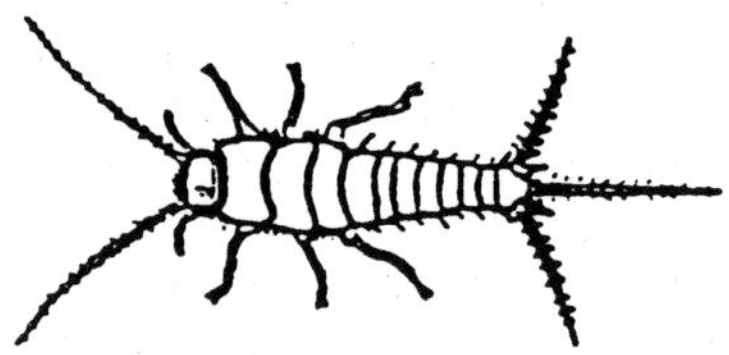

Figure 1.9. Silverfish.

Collembola: Springtails

Springtails (Fig.1.10) are small, wingless insects found in damp habitats such as small pools or damp leaves. Members of this order are characterized by a ventral, spring-like apparatus which allows them to jump considerable distances. Mouthparts are adapted for chewing. Some species occasionally cause damage to gardens by riddling foliage.

Figure 1.10. Springtail.

INSECTS WITH GRADUAL METAMORPHOSIS

Order Orthoptera: Cockroaches, Grasshoppers, Crickets and Mantids

The grasshoppers (Fig. 1.11) cockroaches, crickets, praying mantids and other species belonging to this order possess chewing type mouthparts. When wings are present, the front pair is thickened and leathery and the hind wings are folded underneath like a fan. Cockroaches have long antennae and legs are modified for running; their bodies are flattened from top to bottom, an adaptation that enables them to hide easily in cracks and crevices. Cockroaches are frequently considered important as mechanical carriers of certain enteric diseases of man. Other members of this order, such as the grasshoppers or locusts, constitute one of the major threats to food production in the Near East and Africa. Some species, such as the mantids, are beneficial as predators of insect pests.

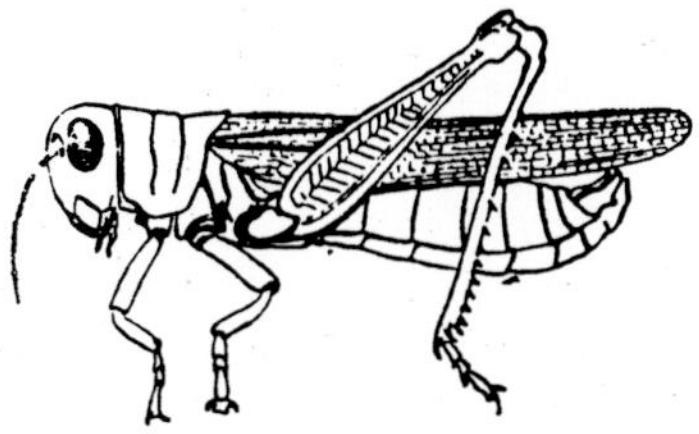

Figure 1.11. Grasshopper.

Order Isoptera: Termites

Termites (Fig. 1.12) are colonial insects that live in concealed nests or colonies as do ants; however, they may be distinguished from the latter in that members of this order have a broad connection between the thorax and abdomen, and winged specimens possess four wings identical in size and appearance. The abdomen is joined to the thorax by a narrow stalk or waist in ants, and the hind wings are smaller than the front pair and different in shape in winged individuals. Unlike ants, bees and wasps, in which the workers and soldiers are exclusively females, termite workers and soldiers exist as both males and females; reproductive organs, however, are undeveloped and they neither mate nor lay eggs. Subterranean termites live in the soil and enter houses by means of protective tubes which extend upward on or through the foundation to the wooden structure itself.

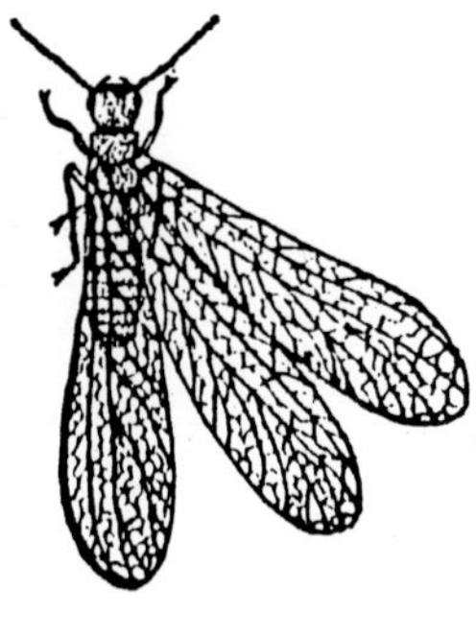

Figure 1.12. Termite.

Order Dermaptera: Earwigs

Earwigs (Fig. 1.13) are rather small, dark-colored insects with a long, narrow body and a prominent pair of forceps-like appendages at the posterior end of the abdomen. Forewings are short, leathery structures which protect the delicate semicircular hind wings. In spite of their foreboding appearance, earwigs are practically harmless; mouthparts are adapted for chewing and most species feed primarily on decaying plant materials although some are carnivorous. In some areas they may invade houses and, like cockroaches, become a very real nuisance.

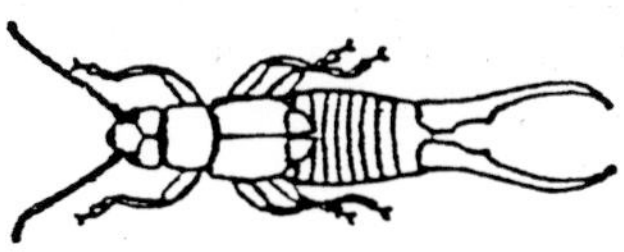

Figure 1.13. Earwig.

Embioptera: Webspinners

Embioptera (Fig. 1.14) are a small, rarely encountered group of insects. Adults usually possess two pairs of wings; however, some species are wingless. Mouthparts are of the chewing type; they feed on plant materials such as fungi, moss, lichens, leaves, dead grass and bark. These are gregarious insects which live in silken tunnels underneath stones, logs and similar environments.

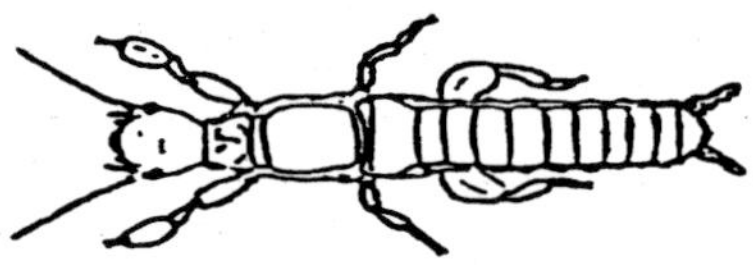

Figure 1.14. Webspinner.

Psocoptera: Psocids and Booklice

Psocids (Fig. 1.15) are small insects possessing chewing type mouthparts and two pairs of wings, usually projecting beyond the end of the abdomen. They may be found on trunks of trees, rocks and log fences where they feed primarily on lichens. Booklice are generally wingless; both nymphs and adults have chewing mouthparts. Most species are general feeders on a wide range of dead and decaying organic matter; they commonly occur among old books in damp darkened rooms that are seldom used.

Figure 1.15. Psocid.

Zoraptera: Zorapterans

Zorapterans (Fig. 1.16) are a very small and rare group of insects. Both winged and wingless forms occur in the same colony; winged forms possess two pairs of wings. Mouthparts are adapted for chewing, and they feed as scavengers or possibly predators on small arthropods. Zorapterans occur under bark, rotten wood and other similar habitats.

Figure 1.16. Zorapteran.

Order Mallophaga: Biting Lice

Biting lice (Fig. 1.17) are external parasites, chiefly of birds (hence the common name, "bird lice"), although some species are parasitic on mammals. They are wingless and possess chewing type mouthparts for feeding on hair and skin scales. Eggs are attached to feathers or hairs; young, termed nymphs, and adults spend their entire lives on the host animal. They infest both domestic and wild animals and birds, causing considerable irritation to the host. At least one species may serve as an intermediate host of a common tapeworm of dogs.

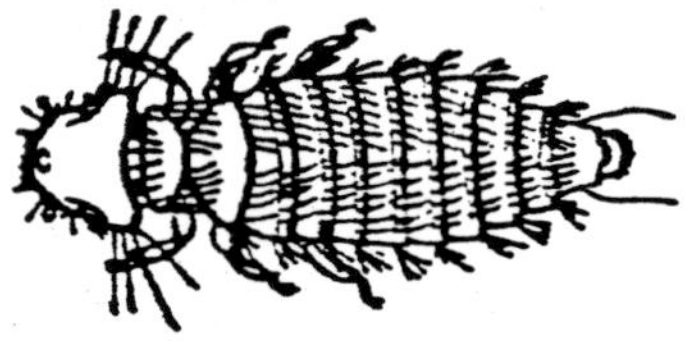

Figure 1.17. Biting louse.

Order Anoplura: Sucking Lice

The sucking lice (Fig. 1.18) are small, wingless, flattened insects that are blood-sucking parasites exclusively of mammals. Mouthparts, which are usually retracted within the head, are modified for piercing and sucking. The eggs, or nits, are attached to the body hairs of the host. Sucking lice cause intense irritation resulting in the host scratching or rubbing against fence posts and trees. Severe infestations may result in production losses and anemia.

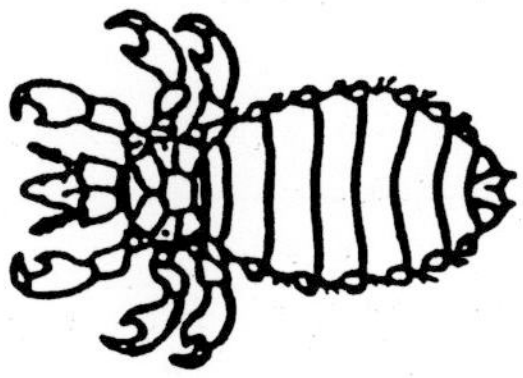

Figure 1.18. Sucking louse.

Thysanoptera: Thrips

Thrips (Fig. 1.19) are small, usually dark colored insects; both winged and wingless forms occur. Wings, when present, are narrow and characteristically fringed with long hairs. Mouthparts are modified for rasping-sucking. The thrips constitute a large order that includes a considerable number of crop, fruit and vegetable pests, although some species are predaceous on other insects.

Figure 1.19. Thrips.

Order Hemiptera: True Bugs

This large order is made up of the "true bugs" (Fig. 1.20), the only insects properly called by that name. The young or nymphs pass through a series of molts before reaching the adult stage. Adults generally have two pairs of wings. The basal half of the front pair (hemelytra) is thickened and the remainder is membranous; hind wings are entirely membranous. A triangular area, termed the scutellum, is located at the bases of the overlapping forewings. Mouthparts are modified into a beak for piercing and sucking. Most of the true bugs are predaceous on other arthropods or are plant feeders that may serve as vectors of numerous plant diseases. Kissing bugs (Family Reduviidae), however, are blood suckers and transmit the trypanosomes responsible for Chagas' disease to both man and dogs, and bed bugs (Family Cimicidae) may become serious pest of poultry and laboratory animals.

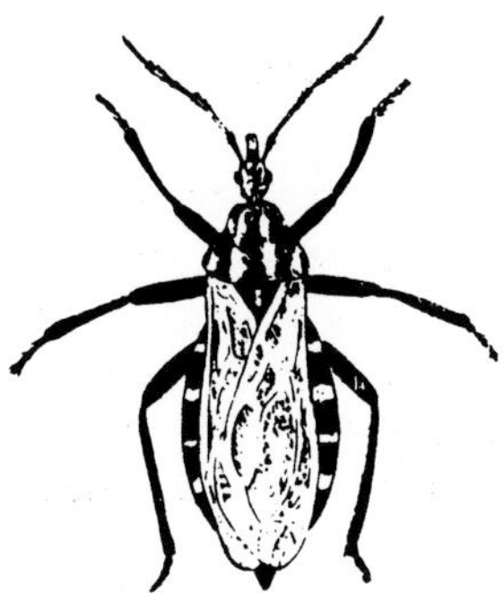

Figure 1.20. Kissing bug.

Homoptera: Cicadas, Leafhoppers, Psyllids, Whiteflies, Aphids and Scale Insects

The Homoptera contain so many different forms that no single common name can be applied to members of this order. Adults have two pairs of wings or are wingless. Wings, when present, form a roof-like cover over the abdomen. (Fig. 1.21). Both adults and nymphs possess piercing-sucking mouthparts. All members of this order are plant feeders.

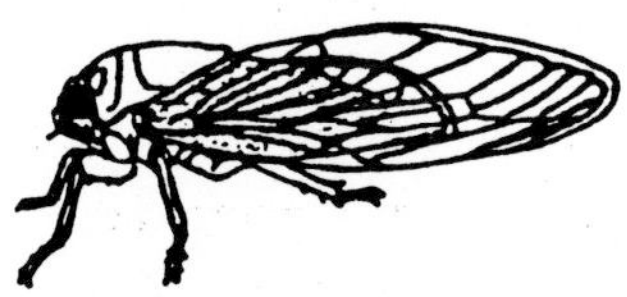

Figure 1.21. Cicada.

INSECTS WITH INCOMPLETE METAMORPHOSIS

Ephemeroptera: Mayflies

Adult mayflies (Fig. 1.22) possess two pairs of membranous wings, characterized by a fine network of veins, and rudimentary mouthparts incapable of feeding. Nymphs are aquatic and complete development to the adult within a few months up to as long as three years. A sub-adult stage, having functional wings, occurs and lasts only a few hours before molting to the true adult which mates and lays eggs. Nymphs constitute an important source of food for fish.

Figure 1.22. Mayfly.

Odonata: Dragonflies and Damselfiies

The Odonata (Fig. 1.23) have two pairs of membranous wings of about equal size. Each wing has a notch on the anterior margin, a character from which the order name is derived. The order is divided into two groups, the dragonflies and damselflies; dragonflies are the larger. Nymphs are aquatic and development requires from less than one year to over three years. Both adults and nymphs possess chewing mouthparts and are predaceous on other insect species in their respective environments.

Figure 1.23. Dragonfly.

Plecoptera: Stoneflies

Stoneflies (Fig. 1.24) possess two pairs of membranous wings, of which the hind pair are larger than the front. Some adults have mouthparts adapted for chewing and feed on algae, but most species possess rudimentary mouthparts and are incapable of feeding. Nymphs, which are aquatic and occur primarily in rapidly running water, have chewing mouthparts and feed on vegetative matter; they constitute an important source of food for fresh water fish. Development requires from one to three years.

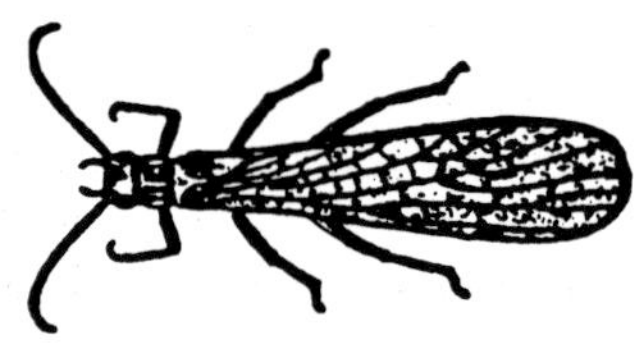

Figure 1.24. Stonefly.

INSECTS WITH COMPLETE METAMORPHOSIS

Neuroptera: Lacewings, Dobsonflies, Ant Lions and Others

The Neuroptera (Fig. 1.25) are characterized by two pairs of membranous wings with numerous veins. Adult mouthparts are adapted for chewing. The biology of the group is quite variable with most larvae being predaceous or parasitic on other arthropods; many species are important in the biological control of insect crop pests.

Figure 1.25. Lacewing.

Order Coleoptera: Beetles

This is the largest order in terms of numbers of species and includes many of the most serious crop pests. In the adult stage, beetles (Fig. 1.26) possess four wings; the front pair, termed elytra, are heavy and shell-like and serve as protective covers for the membranous hind wings. Mouthparts are adapted for chewing. Some species may serve as intermediate hosts of parasitic helminths of wild and domestic animals. Contact with certain vesicating or "blister" beetles (Family Meloidae) may produce severe blistering of the skin. Death of livestock, especially horses, may result from poisoning after animals have eaten hay containing these beetles. The toxic chemical, cantharidin, is a very stable substance that results in severe blistering of the internal organs.

Figure 1.26. Striped blister beetle.

Strepsiptera: Twisted Winged Parasites

The stylopids (Fig. 1.27) are small insects living as internal parasites of other insects. Larval nourishment is taken from the body of the host by osmosis; mouthparts in the adult stage are degenerate or vestigial. Males are winged (four wings); however, the female is wormlike with no eyes or appendages.

Figure 1.27. Strepsipteran.

Mecoptera: Scorpionflies

This is a small order, the most characteristic features of which are the shape of the head, which is prolonged downward into a beak-like structure bearing chewing mouthparts at the tip, and a pincer-shaped organ, which resembles the stinger of scorpions, on the end of the abdomen of males (Fig. 1.28). Adults have two pairs of wings. Both adults and larvae are scavengers, feeding on dead insects and other animal matter.

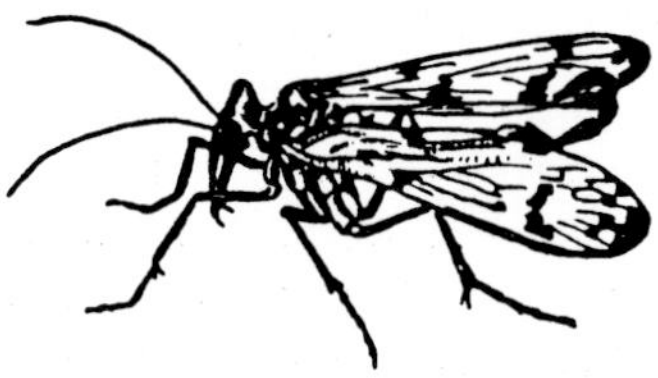

Figure 1.28. Scorpionfly.

Trichoptera: Caddisflies

Adult caddisflies (Fig. 1.29) have rudimentary mouthparts and two pairs of wings that are covered with fine hairs. Larvae are aquatic and live in cases; they possess chewing mouthparts and are largely vegetarian, although some species are carnivorous feeding on other small aquatic arthropods. Caddisfly larvae serve as an important source of food for fish.

Figure 1.29. Caddisfly.

Order Lepidoptera: Moths, Skippers and Butterflies

The moths (Fig. 1.30), skippers and butterflies comprise the second largest insect order. Adults are distinguished from all other insects by large, showy wings that are usually covered with very small colored scales which rub off as a fine powder when specimens are handled. Mouthparts of the adult are of the sucking type, resembling a coiled spring held beneath the head. Larvae, termed "caterpillars", have chewing mouthparts. The larvae of some species possess stinging hairs which may produce severe contact dermatitis to man and animals.

Figure 1.30. Butterfly.

Order Diptera: Flies

The flies (Fig. 1.31), mosquitoes, gnats and midges make up this large order of two-winged insects. A pair of knobs, termed halteres, occur in place of the second pair of wings. Some forms, such as the sheep ked, are wingless. Mouthparts of the adults are modified for sucking or lapping; however, larvae, usually termed "maggots", typically possess chewing mouthparts. Many authorities consider this to be the most important order to medical/veterinary entomology. Mosquitoes, for example, are vectors of dog heartworm and equine encephalitis to animals and malaira, yellow fever and dengue to man. Other species of flies transmit pathogens responsible for bovine mastitis, pinkeye, anaplasmosis and equine infectious anemia.

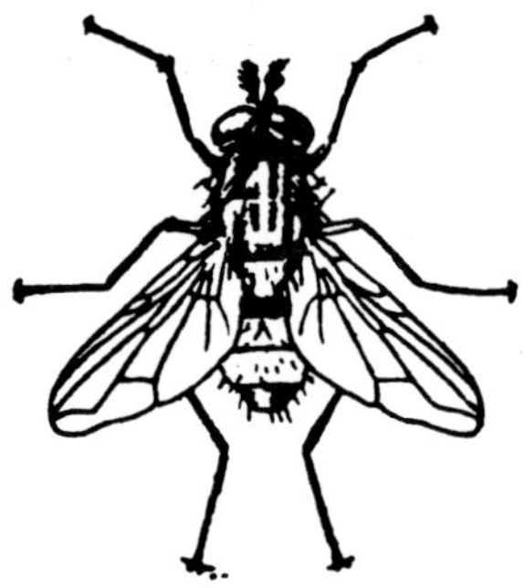

Figure 1.31. House fly.

Order Siphonaptera: Fleas

Fleas (Fig. 1.32) are wingless, blood-sucking ectoparasites of birds and mammals. The body is compressed laterally so that they are quite narrow from side to side. The head bears piercing-sucking mouthparts and antennae; eyes may, or may not, be present. Legs are well developed and adapted for jumping. Many species have black, comb-like structures, termed ctenidia, on the thorax and head. Fleas are important due to their irritating bites and their role as vectors of plague. Several species serve as intermediate hosts for a common tapeworm of dogs.

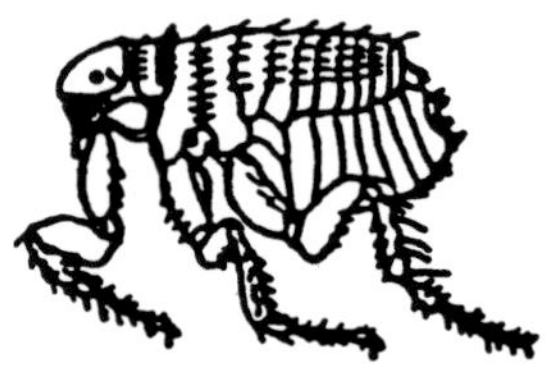

Figure 1.32. Flea.

Order Hymenoptera: Ants, Bees and Wasps

Ants, bees and wasps (Fig. 1.33) make up this large order which ranks third in size in the class Insecta. Hymenoptera are distinguished from most other insects by their four small membranous wings with few veins; hind wings are smaller than the front pair. Mouthparts are modified for chewing or lapping. Many species have the apparent first segment of the abdomen reduced to a slender waist as in the ants. The Hymenoptera is probably the most beneficial insect order in that it contains a great many species that are predators and parasites of many injurious insect pests. However, a number of species can produce painful stings to both man and animals. Some individuals are allergic to these stings and, if stung previously, life-threatening anaphylactic shock may occur as a result of the reaction to subsequent stings by the same species.

Figure 1.33. Honey bee.

CHAPTER 2
MORPHOLOGY AND PHYSIOLOGY

THE EXTERNAL STRUCTURE OF INSECTS

One of the main characteristics of insects and other arthropods is an outer body covering that is hardened into a stiff outer exoskeleton or cuticle. This exoskeleton protects the internal organs from injury and serves as a framework for the attachment of muscles; it also aids in the prevention of water loss from the body. By contrast the skeleton of man and other vertebrate animals is internal and is termed an endoskeleton.

The outer layers of the body wall are hardened or sclerotized into plates, termed sclerites, which are not flexible. These sclerites are joined by flexible portions of the body wall, called intersegmental membranes, which permit considerable movement, as for example when the abdomen of a mosquito becomes greatly distended during feeding. The sclerites may be covered with many small structures, such as hairs, scales, protuberances and spines, many of which are useful for identification purposes.

The body of an insect is divided into three main regions: the head, thorax and abdomen (Fig. 2.1). In the related class Arachnida, which includes ticks, mites, scorpions and spiders, the body is composed of only one or two main regions.

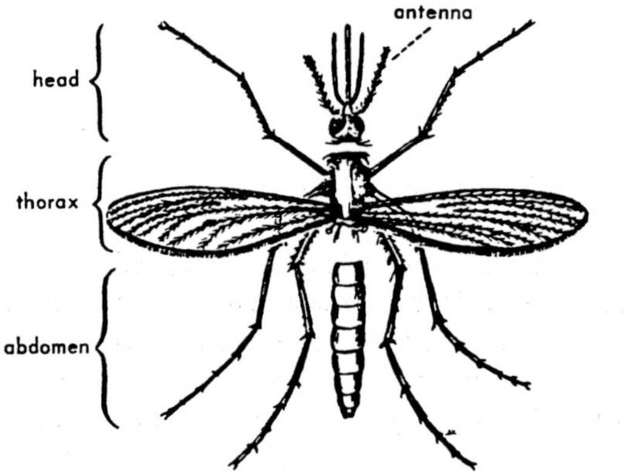

Figure 2.1. Insect body regions. (CDC)

HEAD

The anterior, or first body region, of an insect is the head. Its principal appendages (Fig 2.2) are the mouthparts, the antennae, the large compound eyes, and the simple eyes or ocelli. Adult insects possess only one pair of antennae, whereas arachnids lack antennae and crustaceans have two pairs of antennae.

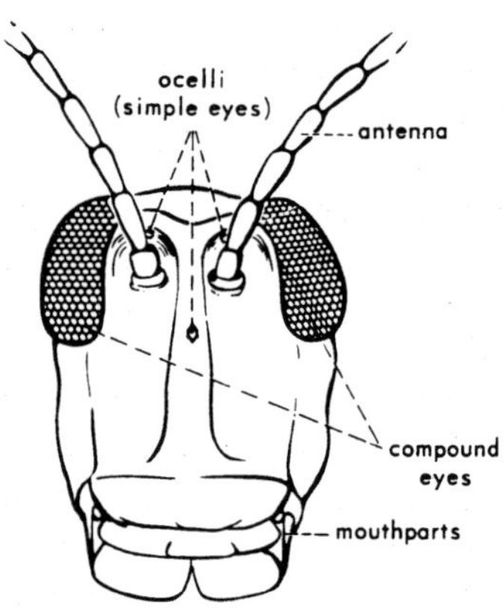

Figure 2.2. Insect head. (CDC)

Mouthparts

The mouthparts of insects consist of an upper lip, or labrum, two mandibles, two maxillae, a lower lip, or labium, and a tongue-like structure termed the hypopharynx. Appendages known as the maxillary palps and labial palps assist in the feeding process and possess sensory receptors used to taste and smell food. The arrangements and forms of these components are extremely variable as a result of adaptations to specific types of feeding. The principal mouthpart types are mandibulate, or chewing, and haustellate, or sucking.

Insects that feed on solid food materials are characterized by chewing mouthparts (Fig. 2.3). The mandibles are the primary cutting and grinding components while the maxillae, labrum and labium aid in the manipulation of food prior to ingestion. The anterior labrum, lateral mandibles and maxillae, and posterior labium form a preoral cavity in which chewing occurs. The tongue-lilke hypopharynx bears sensory receptors and also contains the opening for the salivary duct.

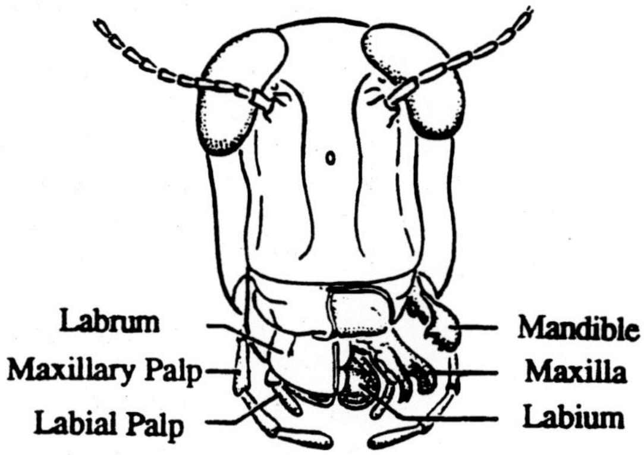

Figure 2.3. Chewing mouthparts. (CDC)

Sucking mouthparts evolved from the more primitive chewing type by a process of elongation of some or all of the mouthpart appendages into a beak-like structure termed the proboscis. Haustellate mouthparts are extremely diverse; however, those species attacking man and animals are typically of two principal types: sponging and piercing-sucking. Sponging mouthparts are adapted for sucking up liquids or readily soluble foods (Fig. 2.4). The mandibles are absent and the maxillae are represented only by their palps; the labrum and labium are joined to form a proboscis with a spongy tip termed the labellum. During feeding, the proboscis is thrust into the liquid or semiliquid food source; capillary grooves on the face of the labellum transport the liquids to a food canal inside the proboscis. Solid foods, such as sugar, that are readily soluble may also be consumed by regurgitation of a drop of saliva from the salivary canal onto the food causing it to dissolve; the solution is then pumped into the mouth as a liquid. House flies, blow flies and flesh flies possess sponging mouthparts and are therefore unable to bite. Many of the blood-sucking vectors of disease possess piercing-sucking mouthparts, which are all modifications of a basic type (Fig. 2.5). In most species, the mouthparts are modified to form a slender beak, or proboscis, which encloses a variable number of stylets; sucking lice do not have a proboscis, but the stylets are retracted within the head. These mouthparts are characteristic of mosquitoes, stable flies, sucking lice, fleas and kissing bugs.

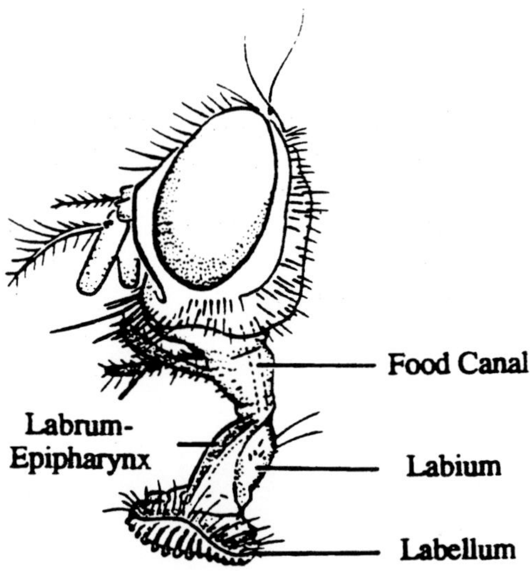

Figure 2.4. Sponging mouthparts. (CDC)

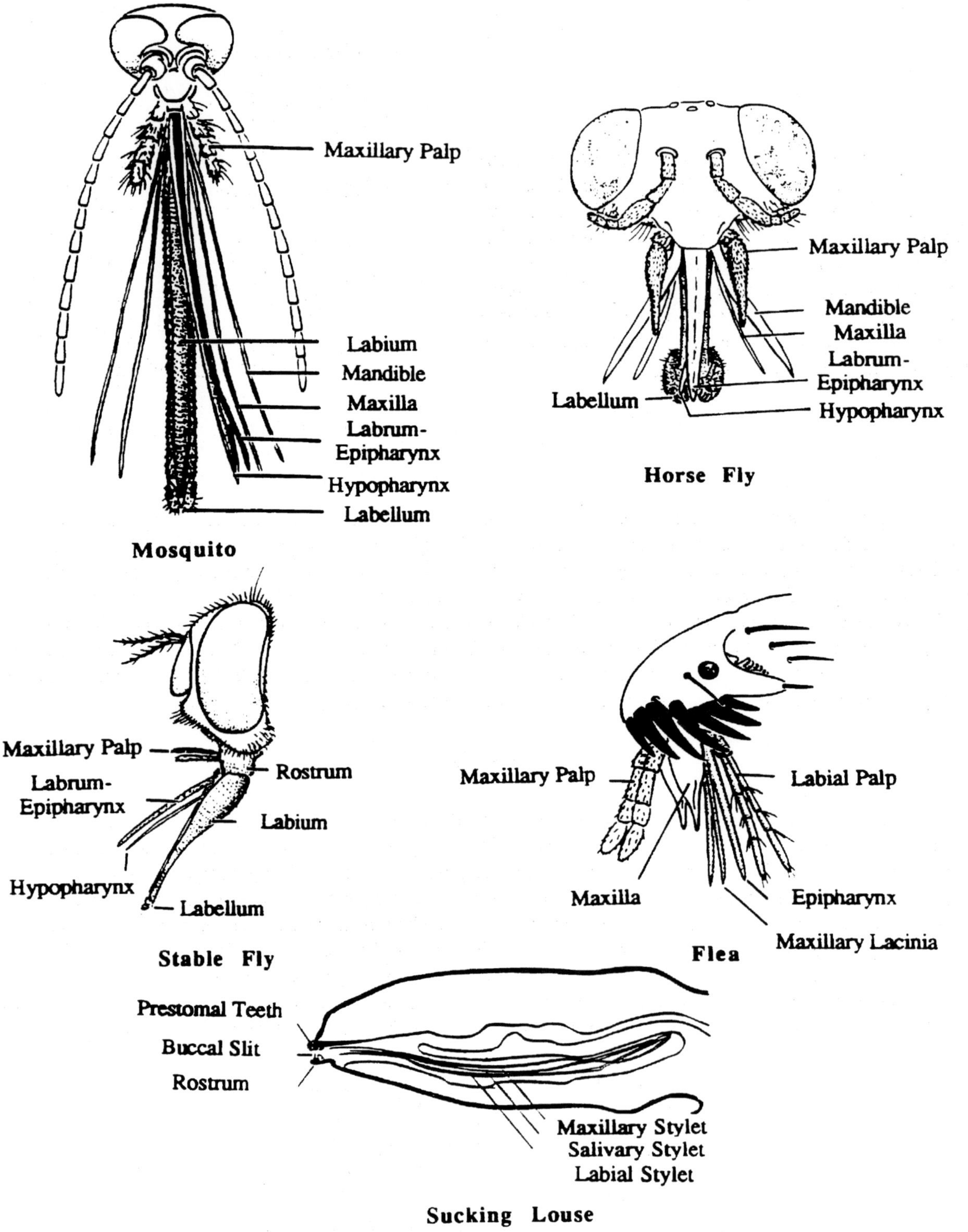

Figure 2.5. Examples of piercing-sucking mouthparts. Clockwise, from upper left: mosquito, horse fly, flea, sucking louse and stable fly. (Redrawn after various authors)

Antennae

True insects, millipedes and centipedes have one pair of antennae situated on the front portion of the head (Fig. 2.2). Most members of the class Crustacea have two pairs of antennae while arachnids have none. The more primitive insects, such as silverfish and cockroaches, have conspicuous antennae composed of numerous, similar segments. In many insects, however, the antennae are greatly modified and often have characteristic shapes that are useful in identification (Fig. 2.6).

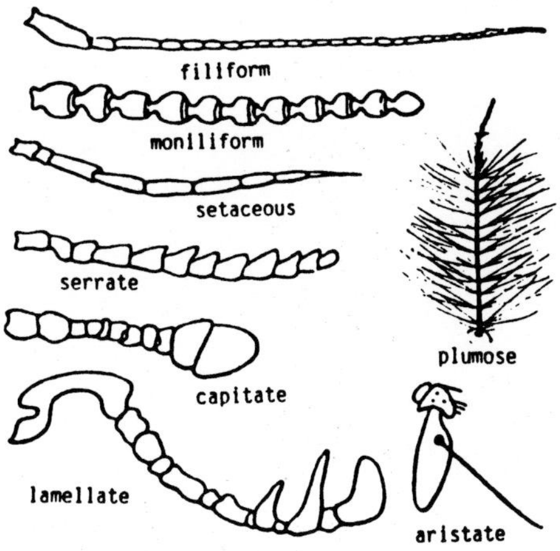

Figure 2.6. Insect antennal types. (CDC)

Eyes

Most insects possess two types of eyes: simple and compound (Fig. 2.2). The simple eyes, or ocelli, consist of single eye units or facets, that are sensitive to light but are not capable of image formation. In many species, three ocelli are arranged in the form of a triangle between the large compound eyes. The compound eyes are usually very large and are round, oval or kidney-shaped. The outer face of the compound eye is composed of many small, six-sided lenses or facets termed ommatidia. Compound eyes are capable of image formation. The size of the eye appears to be related to the need for accurate vision by the insect. In general, active flying insects have large eyes with many facets while more terrestrial types, such as beetles and ants, have fewer facets. Compound eyes are poorly developed in many parasitic insects or, as in certain lice and fleas, may be totally absent.

THORAX

The insect thorax is the second major body region (Fig. 2.1). It is connected to the head by a membranous region, the neck or cervix. The thorax is composed of three segments designated the prothorax, mesothorax and metathorax to indicate their respective positions. Each segment is made up of varying numbers of plates, termed sclerites (Fig. 2.7), with names that are useful in identification. The plate on the top or dorsal side of each thoracic segment is called the notum; on each side, a pleuron; and on the bottom or venual side, the sternum. In identifying insects, the shape of color pattern of the various parts of the thorax is frequently used. For example, in cockroaches, the large oval plate covering the head is the notum of the prothorax, or pronotum, while in mosquitoes, the large plate between the wings is the notum of the mesothorax, or mesonotum. Each thoracic segment bears one pair of legs. Wings, when present, are attached to the mesothorax and metathorax; the prothorax never bears wings.

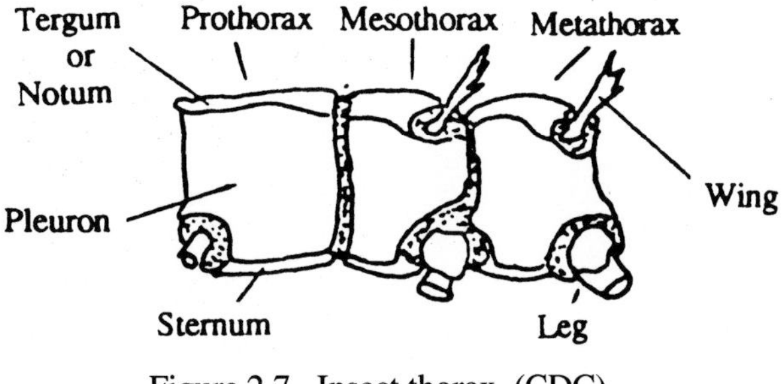

Figure 2.7. Insect thorax. (CDC)

Wings

Insect wings are membranous extensions of the body wall, having an upper and a lower layer supported by reinforcing structures termed veins. Wing veins running from the base to the apex of the wing are referred to as longitudinal veins; cross veins run crosswise to the wing and connect the longitudinal veins. The arrangement and number of wing veins (Fig. 2.8) are important characters for identification of many insects.

Insects typically have two pairs of wings; however, in certain groups, such as the flies, the second pair is modified into small knobs or balancing organs termed halteres. Insects possessing two pairs of wings may use them independently or they may be coupled together as in some of the Lepidoptera and Hymenoptera. The forewings in certain orders of insects (Orthoptera, Hemiptera and Coleoptera) are modified as protective covers for the membranous hind pair.

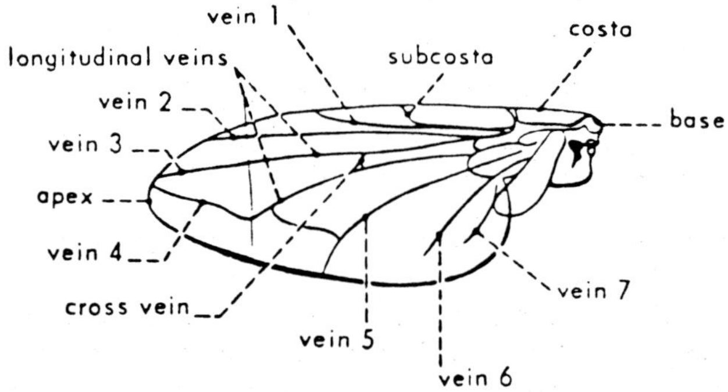

Figure 2.8. Insect wing. (CDC)

Legs

The insect leg is divided into a coxa, trochanter, femur, tibia and tarsus (Fig. 2.9). The femur and tibia correspond to the human thigh and shin, respectively, while the tarsi function similarly to feet. Some of the tarsal segments bear pads, or pulvilli, and claws which assist the insect in walking on smooth surfaces. Figure 2.9 illustrates some of the many types of insect legs. The legs of lice are short and stout and modified into claw-like structures for grasping and holding onto hairs of the host. Legs of fleas and cockroaches are well developed and elongated for jumping and running, respectively.

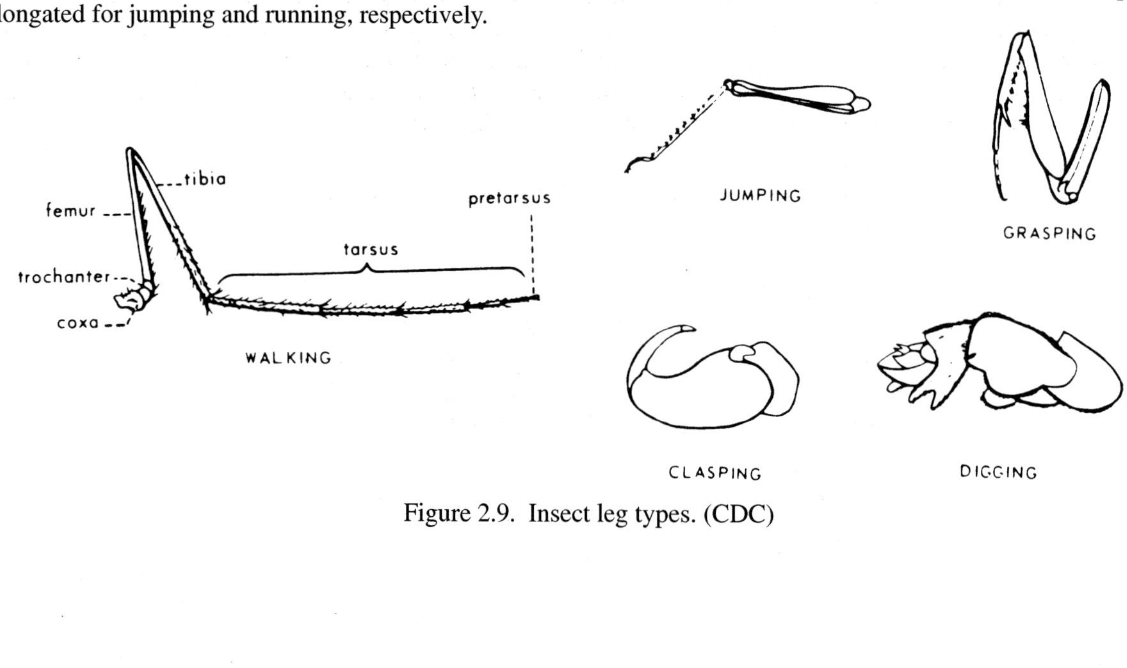

Figure 2.9. Insect leg types. (CDC)

ABDOMEN

The abdomen, or third body region, (Figs. 2.1, 2.10) is made up of segments with dorsal tergites and ventral sternites. It bears the spiracles and the external organs of reproduction. The spiracles are the external openings of the respiratory system. In most insects, the last segments are modified into external sex organs for copulation and, in the female, into an egg-laying apparatus termed the ovipositor. Some insects also bear a pair of terminal appendages termed cerci (singular: cercus).

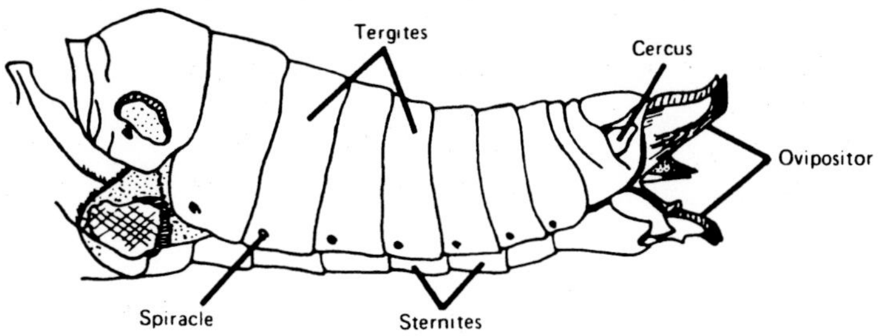

Figure 2.10. Insect abdomen. (CDC)

THE INTERNAL STRUCTURE AND PHYSIOLOCY OF INSECTS

Insects and man differ not only in their skeletal system, the insect having an external skeleton (exoskeleton) and man having an internal skeleton (endoskeleton), but also in the arrangement of the major internal organ systems. The positions of the "heart", or chief circulatory organ, and the nervous system are the exact opposite in insects and man. In the insect the "heart" is on the dorsal, or upper side, and the central nerve cord is ventral, or underneath. In man, by contrast, the heart is on the ventral side and the main nerve cord is encased in the spinal column on the back, or dorsal side. Moreover, the insect circulatory system is an "open" type with one tubular vessel for pumping the blood into the head region from which it flows backward through an open, sac-like body cavity. In man, by contrast, the blood flows through a "closed" system of vessels to and from the heart.

DIGESTIVE SYSTEM

Insects feed on a variety of foods, such as blood, animal tissues, plant tissues, plant juices and wood. The mouthparts and digestive systems are adapted to these specific feeding habits. The digestive system is basically a tube running from the mouth to the anus. It is divided into the stomadeum, or foregut; mesenteron, or midgut; and proctadeum, or hindgut (Fig. 2.11). Food taken into the mouth passes through the esophagus to the crop, which is an enlarged portion of the foregut used for storage. Some insects possess a proventriculus, or gizzard, where food may be ground into finer particles. The food then passes into the stomach where digestion and absorption occurs. Undigested residues, pass out through the intestine and are eliminated through the anus as feces. Most insects possess salivary glands and gastric caecae that secrete enzymes for digestion.

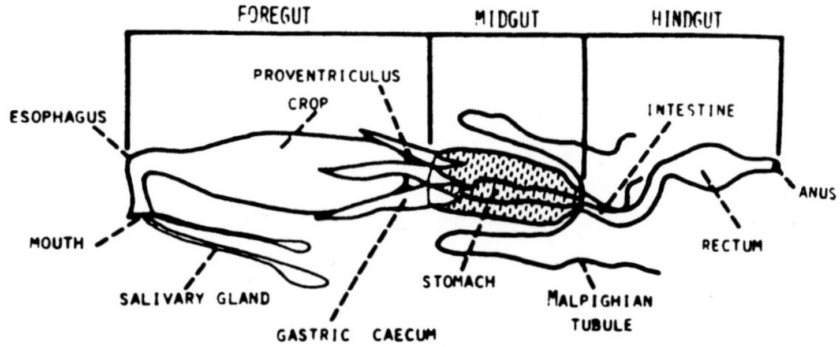

Figure 2.11. Insect digestive system. (CDC)

The digestive system plays a key role in the transmission of many arthropod-borne diseases since the arthropod must pick up the disease-causing organism from one host and pass it on to another. In its simplest form, this transfer simply may be contamination of the arthropod mouthparts with pathogens. For example, if a horse infected with the virus of equine infectious anemia brushes off a stable fly during the process of feeding, virus particles may remain on the fly's mouthparts; if the stable fly returns to feed on a nearby host, these virus particles may be transferred from its contaminated mouthparts to the second animal. This type of transmission is termed mechanical disease transmission by interrupted feeding. Similarly, feeding house flies may suck up enteric pathogens from filth; later, when they regurgitate saliva to dissolve sugary or starchy foodstuffs, they may transfer the pathogens to human foods. Pathogens may also pass through the digestive system of flies and be introduced into foodstuffs via fly feces.

The digestive system is also involved in many arthropod-borne diseases involving complex life cycles since the vectors of major importance are usually blood-sucking species. Disease organisms are ingested along with blood from an infected host and are then transferred to other hosts during subsequent feedings. Mosquitoes may feed on a host whose blood contains viruses such as those of equine encephalitis. These viruses pass out of the mosquito's digestive tract into its body cavity, later develop in the salivary glands and are injected with salivary secretions when the mosquito subsequently feeds on another host. With plague, a flea feeding on an infected rodent or human host ingests plague bacteria along with blood. In the flea's body, these bacteria multiply very rapidly so that the proventriculus and the ventriculus become completely filled and blood can no longer pass. The hungry, blocked" flea continues attempting to feed, often on another host. Blood drawn up as far as the proventriculus cannot pass and is regurgitated, carrying with it plague bacilli that may infect the new host's body through the bite wound. Some other disease organisms, for example the trypanosomes of Chagas' disease in the kissing bug, develop in the insect gut and are voided with the feces. As the infected arthropod feeds, sucking up more and more blood, the older, partly digested blood containing the disease organisms is forced out the hindgut in droplets of liquid feces. When a dog bitten by an infected arthropod scratches the bite wound, the infected feces are rubbed into the skin and infection results.

EXCRETORY SYSTEM

An arthropod's undigested food is expelled from the body as feces whereas the excretions of true body wastes, the by-products of growth and metabolism, are transported by the blood to the Malpighian tubules where they are discharged into the intestine at the junction of the midgut and the hindgut (Fig. 2.11).

CIRCULATORY SYSTEM

The circulatory system of insects is an open type in the sense that the blood is not enclosed in a system of vessels but circulates freely throughout the body cavity, or haemocoel. The blood passes into the posterior portion of the dorsal blood vessel, or heart, through paired valves (ostia) and is pumped into the head region through the anterior portion termed the aorta (Fig. 2.12). From the head region the blood flows posteriorly, bathing the various body tissues. Small pulsating organs help to circulate the blood in the legs and antennae. Typically, insect blood is either colorless or greenish-yellow. Only in a few insects, such as blood-worm midge larvae (*Chironomus*), does the blood have a reddish color due to the presence of hemoglobin. Insect blood does not serve in the transport of oxygen and carbon dioxide as in mammals. The main functions are transportation of nutrients to body cells for growth and energy production, and of waste products from body cells to the Malpighian tubules for excretion. Insect blood also contains several different types of blood cells which function in phagocytosis and wound healing.

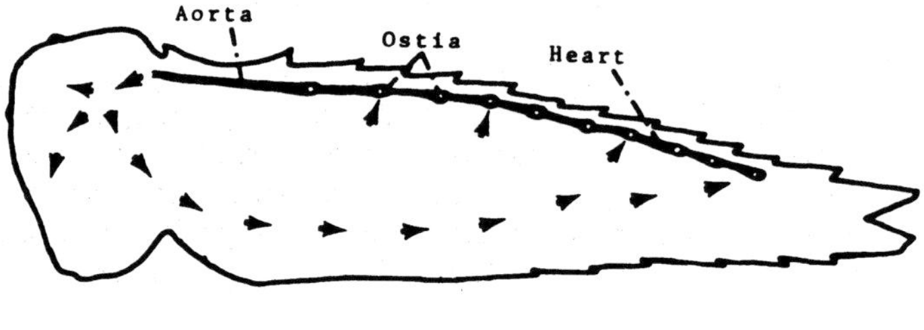

Figure 2.12. Insect circulatory system. (CDC)

NERVOUS SYSTEM

The insect nervous system (Fig. 2.13) contains nerve tissues and organs much like those found in higher animals. The brain lies in the head above the esophagus and is connected to a subesophageal ganglion by two nerve cords encircling the esophagus. A double, ladder-like nerve cord extends posteriorly along the ventral surface of the body cavity. In primitive insects, each body segment is innervated by a nerve center or ganglion (plural: ganglia); however. the number is reduced in most higher forms due to fusion anteriorly.

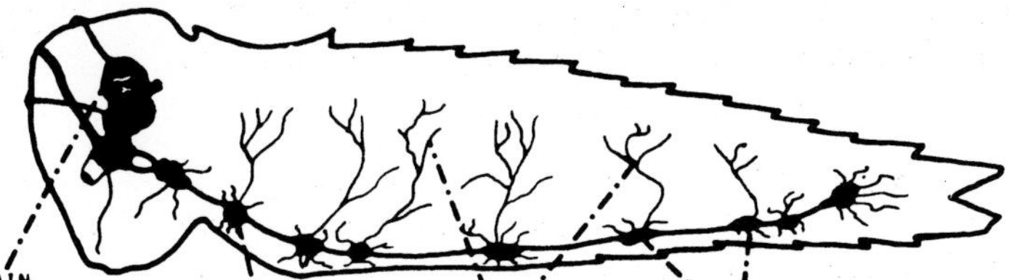

Figure 2.13. Insect nervous system. (CDC)

RESPIRATORY SYSTEM

Air enters the insect body through external openings, termed spiracles, in the body wall into large tracheal trunks (Fig. 2.14) which usually extend the length of the body. Many tracheae branch off these main trunks and carry air to the tissues through finely branched tracheoles. Respiratory movements of the insect body alternately compress and expand the large tracheal trunks thereby ventilating the main branches of the respiratory system. Some insects are able to regulate the flow of air, taking it in through the anterior spiracles and expelling it through posterior spiracles. Oxygen is carried to all parts of the insect body directly by these tracheal tubes. This same system also acts to carry carbon dioxide from the body tissues. Considerable loss of water through this system would occur if it were not for the presence of valves located within the spiracles. In some insects, these valves are regulated by the presence of carbon dioxide, an excess of which causes them to open. Oxygen requirements of insects are of the order of one cubic millimeter of oxygen per gram of body weight per hour, but varies greatly with species and environmental conditions such as temperature and humidity.

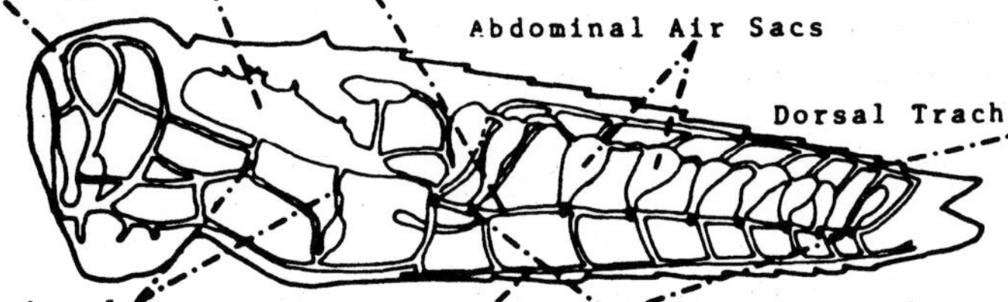

Figure 2.14. Insect respiratory system. (CDC)

REPRODUCTIVE SYSTEM

Most insects are either male or female, and mating must occur before viable eggs are developed. Large numbers of eggs are usually produced, although some species develop very few eggs and others may produce living larvae. Insects which deposit eggs are termed oviparous while those species producing larvae are referred to as larviparous. The sheep ked, for example, gives birth to full-grown larvae that pupate almost immediately. A few insects have only one sex, the female, which is capable of reproducing without fertilization as in the Surinam cockroach. This type of reproduction is called parthenogenesis. Such species are not hermaphroditic as no male organs are present.

The male reproductive system (Fig. 2.15) consists of a pair of testes, in which sperm cells are developed and ducts, or vas deferens, leading to the penis, or ejaculatory organ. The seminal vesicle serves as a reservoir for storing sperm cells until mating occurs. Accessory glands secrete a liquid substance serving as a vehicle for the sperm cells.

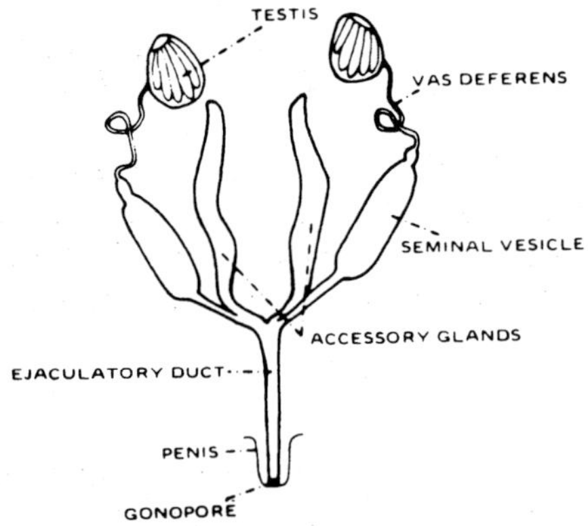

Figure 2.15. Male insect reproductive system. (CDC)

The female reproductive system (Fig. 2.16) consists of a pair of ovaries, which produce eggs (ova), and paired oviducts through which the eggs pass into the vagina where they may be fertilized by male sperm cells stored in the spermatheca (plural: spermathecae). Some species possess accessory glands that secrete an adhesive coating for the eggs. A single copulation usually supplies the female with sufficient sperm to fertilize a large number of eggs whether they are deposited at one time or at intervals over a long period.

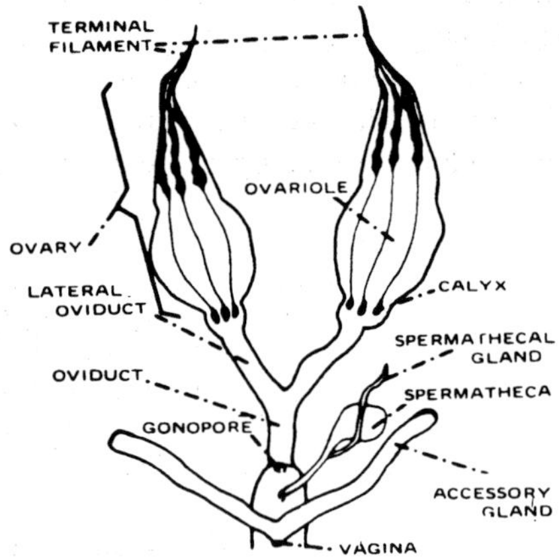

Figure 2.16. Female insect reproductive system.(CDC)

Honeybees and the screwworm fly mate only once. However, many insects mate several times. Some insects, such as the honeybee, can deposit infertile eggs to produce males (drones) and fertile eggs to produce females (workers and queens).

CHAPTER 3
THE DEVELOPMENT OF INSECTS

The life history of an insect begins with the fertilization of the egg and is completed when the adult stage is reached. The term "life span", or "life cycle", refers to the entire length of life of the insect. Some insects, such as tropical termite queens, may survive 15-20 years and the periodical cicada lives for 14-17 years. Adult mayflies, on the other hand, survive only a few days although they may spend 2-3 years as developing immature stages.

MOLTING

Small animals have relatively greater surface areas in proportion to their weight and volume than do large ones. This results in increased evaporation of water vapor from the body and requires the development of a complex waterproof body covering, the exoskeleton or cuticle, which also serves as protective armor as well as body support and as the framewok for muscles of locomotion. However, the development of this armor-like exoskeleton (Fig. 3.1) complicates the growth of insects. The process by which an immature insect sheds this protective skeleton, along with the linings of the respiratory system and of the foregut and hindgut, is termed molting or ecdysis. To accomplish this process, a molting fluid is produced between the old exoskeleton and the new soft cuticle of the insect. Air or water is swallowed until the body swells, bursting the old cuticle on the dorsal surface in most cases. The insect gradually extracts itself from the old exoskeleton and a new cuticle then hardens on its expanded body, thus accomplishing a stage of growth termed an instar. The number of molts may be small and constant, such as the three molts of the house fly, or it may be large and variable, as in the twelve or more molts of the American cockroach. For a period varying from a few hours to a day following molting, the body of the insect may be soft and pale colored; however, progressive hardening and coloring of the new cuticle soon occurs.

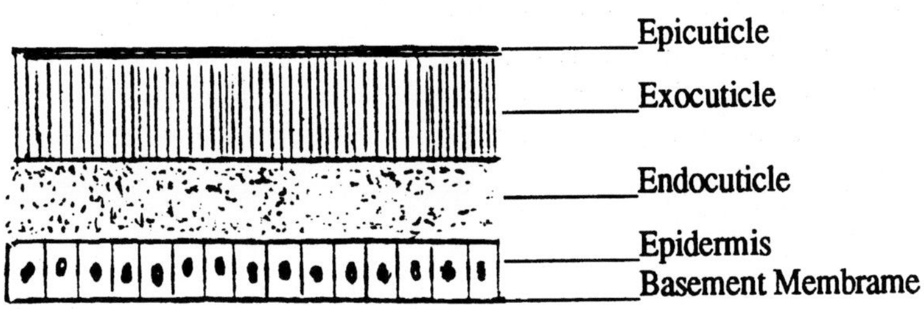

Figure. 3.1. Insect exoskeleton (cuticle).

Molting in insects is under the control of two groups of hormones: (1) the molting hormones, or ecdysones, and (2) the juvenile hormones. Ecdysones are necessary for molting to take place, playing a key role in resorption of the old cuticle and deposition and hardening of the new one. Juvenile hormones must be present at each molt during the immature stage to prevent the insect from maturing; conversely, they must be absent during the last molt to allow the immature stage to mature into the adult insect.

METAMORPHOSIS

Metamorphosis refers to changes in the form or structure of an insect during the course of its development. A few primitive insects develop without metamorphosis (ametabolous development). The young possess all the obvious structures of the adult, differing merely in size, color and sexual maturity. Springtails (Order Collembola) and silverfish (Order Thysanura) develop without metamorphosis. Both are small, wingless insects. Thysanura continue to grow and molt throughout life.

Insects with gradual (paurometabolous development) and incomplete metamorphosis hemimetabolous development) pass through three stages (egg, nymph and adult) (Fig. 3.2) during the course of their development. Insects in this group change gradually while undergoing a succession of molts to become adults. The young, termed nymphs, resemble the adult insect except for their smaller size, sexual immaturity and the absence of wings in winged species; wing pads may be present in the latter stages of development. Nymphs of insects characterized by incomplete metamorphosis are aquatic and are termed naiads.

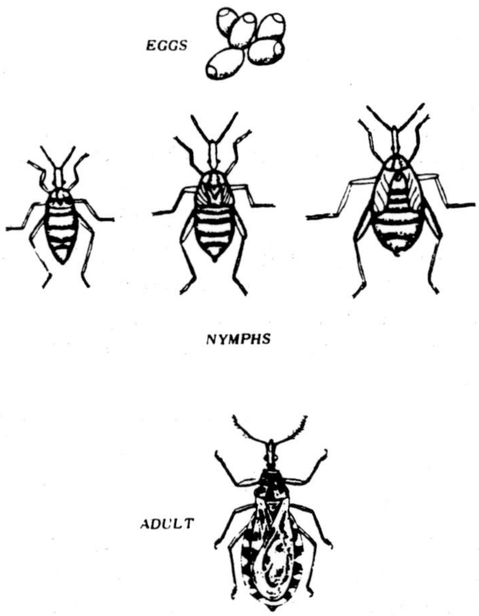

Figure 3.2. Developmental stages of insects with gradual and incomplete metamorphosis. (CDC)

Insects with complete metamorphosis (holometabolous development) undergo four developmental stages (egg, larva, pupa and adult) (Fig. 3.3). Immatures of insects with this type of life history are morphologically distinct from the adult stages. Typical larvae are the wigglers of mosquitoes, the maggots of flies, the grubs of beetles and the caterpillars of butterflies and moths. The pupal stage is an important evolutionary development during which the simple larva undergoes many external and internal changes during the transformation into the adult insect. Most insects with complete metamorphosis possess wings as adults, but some species, such as the fleas, are completely wingless. Normally, wing buds first appear during the pupal stage. When the adult first emerges from the pupal case, the wings are crumpled and incapable of flight; hydrostatic pressure of the blood within the insect body forces the sac-like wings outward and the two membranes collapse against each other to form the single membranous structure.

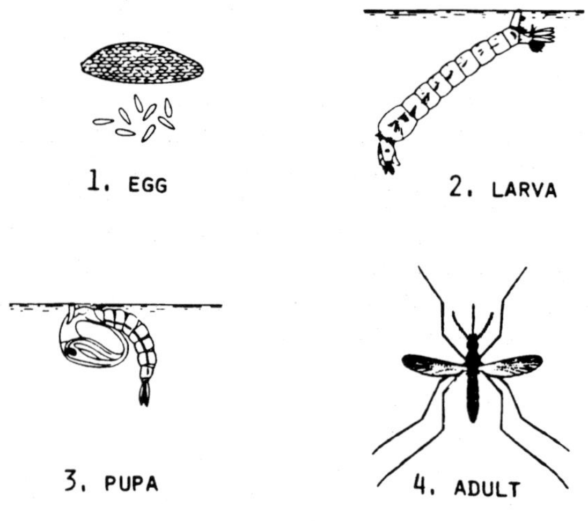

Figure 3.3. Developmental stages of insects with complete metamorphosis. (CDC)

CHAPTER 4
PARASITISM AND DISEASE TRANSMISSION

Arthropods generally affect livestock and poultry in one of two major ways. The effects they produce on a host may be directly attributed to their presence, in which case the arthropod and the host constitute the only two living factors involved in the association. In other instances, however, the arthropod serves only in an indirect capacity by transmitting some pathogen, or disease-producing organism, from one host to another so that three living factors are involved in the association.

HOST-PARASITE INTERACTIONS (TWO LIVING FACTORS)

Parasitism is a type of association between two different animal species in which one, the parasite, is benefited while the other, the host, suffers some degree of physical injury or harm. Most insects, ticks and mites associated with livestock and poultry are ectoparasites situated on the outer body surface of the host. The majority are blood feeders; however, a few, such as the biting lice, feed on feathers and hair. Others, such as the house fly and face fly, feed on body fluids while screwworm larvae feed directly on body tissue itself. The presence of external parasites on a host is termed an infestation. Endoparasites, such as lung worms and stomach worms, are located within the body of the host; their presence is termed an infection.

Parasites may be further classified as obligatory or facultative (occasional) depending upon their degree of reliance on a living host to successfully complete their life cycle. Obligatory parasites, such as lice and fleas, are unable to reproduce in the absence of a host. Some arthropods, such as the house fly and most species of blow flies, are facultative parasites which are capable of completing their life cycles without a host.

Parasites also may be classifted as continuous or intermittent in their association with a host. Lice, for example, are continuous parasites in that they complete their entire life cycle in direct contact with the host. Mosquitoes and many other species of biting flies, on the other hand, are intermittent parasites, coming into contact with the host for only short periods of time in order to feed.

Production losses to livestock and poultry directly attributable to arthropod parasitism are rarely so serious as to result in death of the host. The most common direct effect is annoyance and irritation ("worry"), which results in physiological stress and is frequently manifested in a general unthrifty condition. Even biting lice which feed on hair, feathers and skin debris can produce serious annoyance due to their continual movements over the skin surface. Annoyance coupled with blood loss also may result in anemia, lowered weight gains and decreased milk and egg production. Dermatitis may produce serious damage to wool and hides. The injection of salivary secretions during the process of feeding, as well as certain venomous bites and stings, may result in specific allergic reactions including anaphylactic shock.

HOST-PATHOGEN-VECTOR INTERACTIONS (THREE LIVING FACTORS)

In certain diseases, the arthropod serves in an indirect capacity acting as a living vector transmitting the pathogen, or disease-causing organism, from one host to another. Microbial pathogens most frequently transmitted are viruses, rickettsiae, bacteria and protozoa; arthropods also frequently serve as intermediate hosts of many species of parasitic worms (helminths). Vector-borne diseases vary in complexity, mode of transmission and range of hosts involved. They are often separated into two major groups according to whether transmission is merely mechanical or whether it involves some biological process within the body of the arthropod vector.

Mechanical Transmission of Disease

Mechanical or passive transmission of disease occurs when an arthropod transports pathogens from one host directly to another on its mouthparts, feet, body hairs or other surfaces. There is no multiplication or developmental change of the disease organism within the body of the arthropod. Infected mouthparts resulting from frequent interrupted feeding attempts constitute the most common method of mechanical transmission of diseases to livestock. Subsequent feedings usually must take place within a matter of minutes after the initial attempt for successful transmission to occur. Anaplasmosis and equine infectious anemia commonly are transmitted by horse flies and stable flies in this manner. House flies also are capable of transmitting Newcastle disease to poultry by contaminated mouthparts and body parts.

Biological Transmission of Disease

Biological disease transmission occurs when the arthropod not only transmits the pathogen from one host to another but is essential to the life history of the disease organism. Multiplication and/or a developmental change of the pathogen must occur within the body of the arthropod vector. The period of time during which the pathogen must proliferate and/or develop in the arthropod is referred to as the extrinsic incubation period, while the interval required for the organism to develop and produce disease after injection into a new host is the intrinsic incubation period. Three main types of biological transmission are commonly recognized:

1. Propagative disease transmission, in which the pathogen merely increases in number without undergoing any developmental cycle within the body of the arthropod. A typical example is the virus responsible for equine encephalitis which multiplies in the body of certain species of mosquitoes. After a variable number of days, these viruses concentrate in the salivary glands to be injected as an infective dose when the mosquito subsequently feeds on another host. The development of plague bacteria in the alimentary canal of a flea is another example of this type of disease transmission.
2. Cyclo-developmental disease transmission, in which the pathogen undergoes an essential part of its developmental cycle in the body of the arthropod without increasing in number. With dog heartworm transmission, for example, the mosquito sucks up a number of microfilariae into its alimentary canal during the process of feeding. These forms eventually develop into infective stages within the mosquito's malpighian tubules, after which the infective larvae migrate to the proboscis to infect another host when the mosquito feeds again.
3. Cyclo-propagative disease transmission, in which there is both an increase in number and a change in the developmental cycle of the pathogen in the body of the arthropod. For example, the protozoan parasite responsible for babesiosis or cattle fever undergoes an essential portion of its life cycle and also considerably increases in number within the body of its tick vector.

Transovarian disease transmission constitutes a highly specialized type of biological disease transmission in which the pathogen is passed from one generation of arthropods to the next via the egg through the infected ovaries of the parental female. Transovarian disease transmission has been reported to occur in a number of arthropods, such as mosquitoes; however, this type of disease transmission is the most common manner by which ticks transmit pathogens. Transstadial transmission constitutes another mode of transmission, primarily involving ticks, in which an immature stage having fed on an infected host will obtain the pathogen and transmit it to an uninfected host during a subsequent stage of its developmental cycle.

HOSTS

The reservoir host, defined as one or more hosts or vectors which serve to harbor a pathogen, usually over an extended period of time, is also important to understanding the transmission of vector-borne diseases. Reservoir hosts typically do not show any symptoms of infection. Susceptible hosts, on the other hand, usually show overt, clinical signs of disease and tend to maintain the pathogen for a shorter duration. Diseases in animal reservoirs which smolder at more or less constant levels year after year are referred to as enzootic while an usually high level of disease in animal populations is termed epizootic. The terms endemic and epidemic are usually used to describe these respective states in humans.

CHAPTER 5
LICE

Lice are wingless, dorso-ventrally flattened, obligatory ectoparasites of birds and mammals. Once on the host, they rarely leave voluntarily, except where two hosts are in direct body contact or if the host should die, thereby losing its attractiveness. During its lifetime, a single host may, therefore, support multiple generations of lice and serve as a source of infestation to both its mates and its offspring. Many species of lice grow and reproduce slowly, resulting in only minor discomfort to the host, whereas other, more prolific species explode into enormous populations under optimum conditions. Under these latter circumstances, the debilitating effects of increasing populations may weaken the host's resistance to attack, allowing development of larger and larger numbers of lice.

ORDERS OF LICE

Lice are most frequently placed in two separate orders because of important differences in the structures of their mouth parts. The Mallophaga (biting or chewing lice) have mandibulate mouthparts, located on the ventral side of the head, which are used to feed on feathers, hair and epidermal skin scales. Some species of Mallophaga also feed on blood which they obtain by piercing the pulp of young feathers or by gnawing through the skin. Members of this order may be readily identified by a head that is wider than the prothorax; species parasitizing birds typically have two claws on the distal end of each tarsus while those on mammals usually have only a single claw. There are approximately 3,000 described species of Mallophaga in the world. In contrast, the Anoplura (sucking lice) are characterized by a head width narrower than that of the prothorax. Mouthparts consist of three piercing stylets contained in an eversible pouch concealed within the head (Fig 2.5); feeding is accomplished by everting the pouch and forcing the mouthparts through the skin into a blood vessel. Only about 500 species of Anoplura have been described. A few authorities place all lice in the single order. Phthiraptera, in which Mallophaga and Anoplura are suborders.

HOST SPECIFICITY

With few exceptions, lice are highly host specific. Most species, therefore, can survive only on these hosts and not on others. The factors responsible for this specificity are not entirely understood; some of them are probably physiological in nature. Host specificity developed over a period of thousands of years as lice evolved along with their hosts while being transmitted from parent to offspring.

GENERALIZED LIFE CYCLE

A louse develops by gradual metamorphosis (Fig. 5.1). The oblong eggs, commonly termed nits, are approximately 1 mm long and are deposited several at a time over an interval of 2-3 weeks. Eggs are glued to the hairs or feathers of the host with a secretion from the female accessory glands and remain firmly attached during embryogenesis and hatching. Average egg production varies with the species from 150-300 eggs per female. Hatching usually occurs several days up to three weeks following oviposition. Nymphs physically resemble the adults except for their smaller size and somewhat lighter coloration. All lice develop through three nymphal instars during a period of 2-3 weeks. Nymphs and adults are found at the same location on the host. Adults of most species are light tan to brown in color and range from1-4 mm in length. However, several larger species (5-7 mm) may be found on livestock, and lice as much as 10 mm long occur on some wild birds.

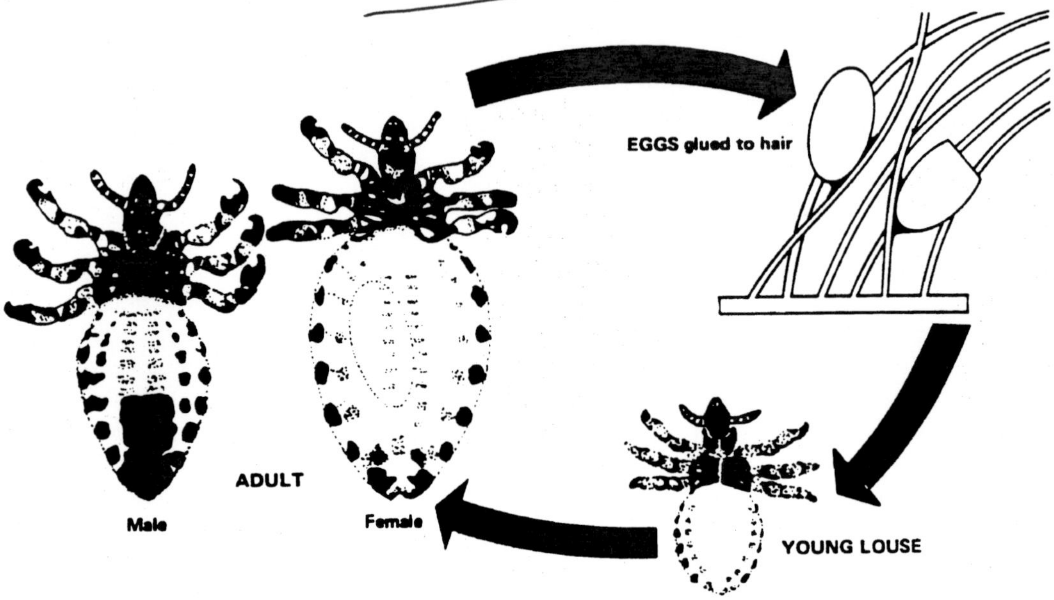

Figure 5.1. Generalized louse life cycle. (CCEX)

POULTRY LICE

Poultry are attacked only by biting lice. Poultry lice are generally more abundant during the summer than in winter, but all stages usually can be found on the host at any time. Generally, these lice do not leave the host's body except to pass from one fowl to another, most commonly from old to young birds. Eggs, which are attached to feathers, are generally white in color and oval in shape. Although deposited singly, eggs may be abundant enough to form dense clusters on the fluffy areas of contour feathers of severely infested hosts. Hatching generally occurs in a few days to a week. Nymphs immediately begin feeding and, during the next few weeks, gradually assume adult size, form and coloration.

The chicken body louse, *Menacanthus stramineus* (Figs. 5.2), which parasitizes domestic chickens and turkeys, is the most common louse of poultry worldwide. Specimens occasionally may be found on domestic ducks and geese, but they do not appear to reproduce on these species. Chicken body lice usually occur where feathers are sparce such as the vent area of the host. Eggs hatch in 4-5 days, with 14 days required for nymphs to mature to adults. The preoviposition period is three days; females deposit an average of 1.5 eggs per day over a 12.5-day interval. The shaft louse, *Menopon gallinae* (Fig. 5.3), is also common on domestic chickens and guinea fowls worldwide. Shaft lice habitually rest on feather shafts but quickly run onto the body when disturbed. The chicken head louse, *Cuclotogaster heterographus* (Fig. 5.4), normally resides in the feathers of the head and neck. Eggs hatch in 5-7 days; three nymphal instars are completed in an additional 25-42 days. The wing louse, *Lipeurus caponis* (Fig. 5.5), is a slender species usually found between barbules on wing and tail feathers and sometimes on back and neck feathers as well. Fluff lice, *Goniocotes gallinae* (Fig. 5.6), are common on domestic chickens worldwide. One of the smallest louse species of poultry, it most often occurs in the fluff at the base of feathers on the sides of the thighs and on the back. The large chicken louse, *Goniodes gigas* (Fig. 5.7), commonly attacks domestic chickens and guinea fowls in North America, Central America, Europe, Australia and Africa It is a large species, up to 5 mm long, usually found on feathers of the body rather those of the extremities, but seldom in large numbers. Eggs hatch in seven days and the life cycle averages one month.

Large turkey lice, *Chelopistes meleagridis* (Fig. 5.8), occur on turkeys among feathers on the neck and breast. The slender turkey louse, *Oxylipeurus polytrapezius*(Fig. 5.9), is found most frequently on the wing feathers of turkeys.

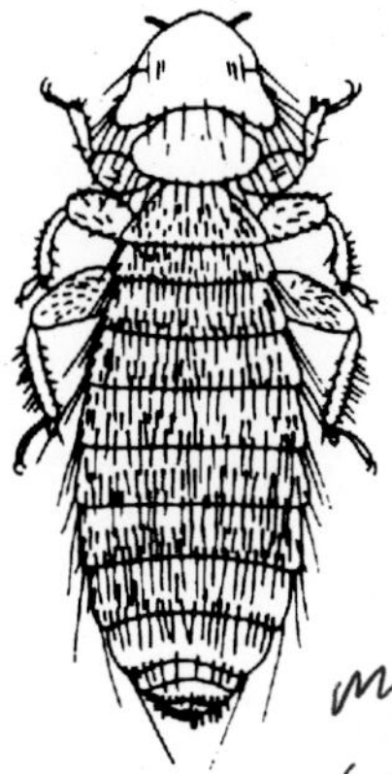

Figure 5.2. Chicken body louse. (TAEX)

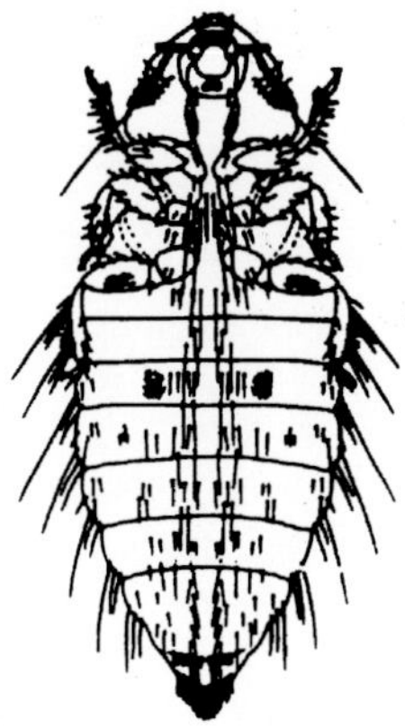

Figure 5.3. Shaft louse. (TAEX)

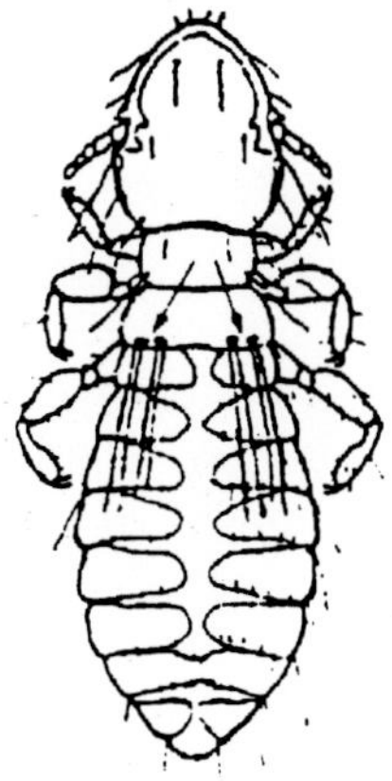

Figure 5.4. Chicken head louse. (TAEX)

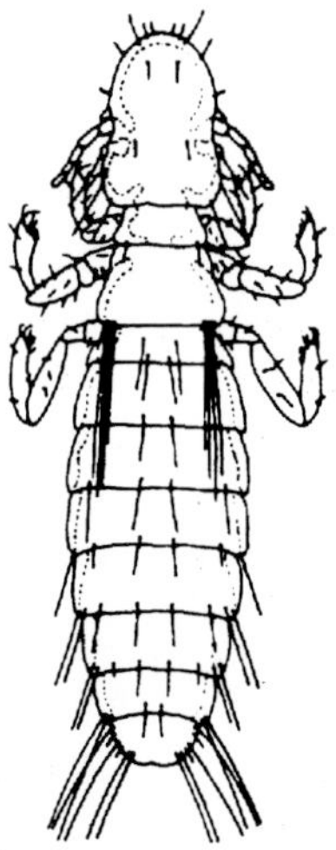

Figure 5.5. Wing louse. (TAEX)

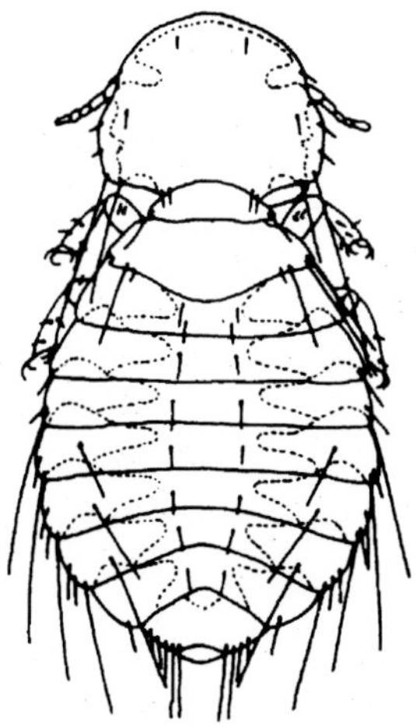

Figure 5.6. Fluff louse. (TAEX)

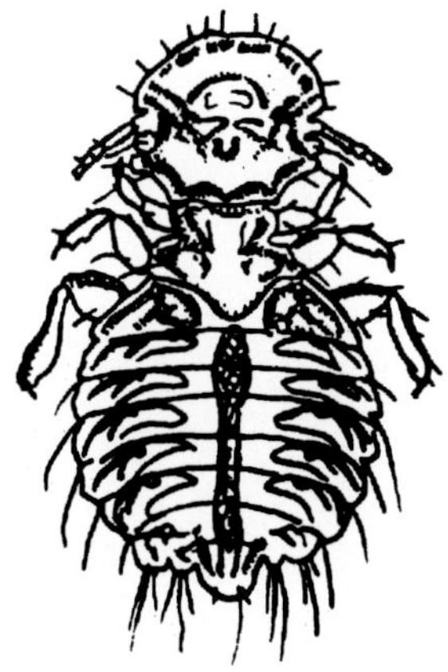

Figure 5.7. Large chicken louse. (TAEX)

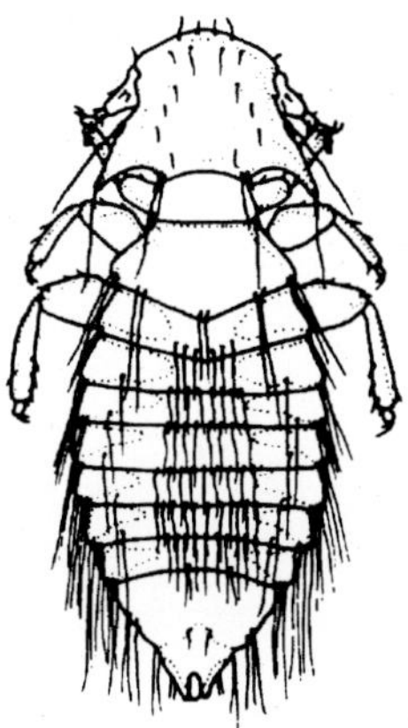

Figure 5.8. Large turkey louse. (TAEX)

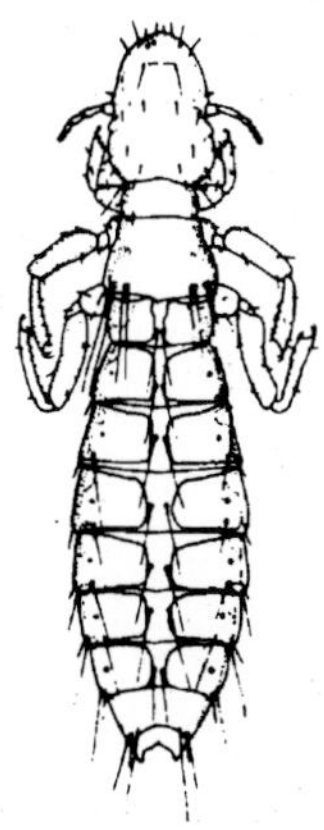

Figure 5.9. Slender turkey louse. (TAEX)

CATTLE LICE

Five species of lice parasitize cattle in North America. The cattle biting louse, *Bovicola bovis* (Fig. 5.10), is readily distinguishable from the four species of sucking lice by its broad head and yellowish-tan abdomen with dark transverse bands. It is a small species, varying in length from about 1-1.5 mm. During the winter, cattle biting lice are found in groups at the base of the tail, on the shoulders and along the back; as infestations increase, the lice spread down the sides to the remainder of the host's body. Individuals persist mainly on the tail, head and perhaps inside the ears during the warm, summer months. The small, light-brown eggs are attached to the hair close to the skin. Under ideal conditions, eggs hatch in 7-10 days, the average being approximately eight days. Time required for completion of the three nymphal stages averages about 19 days. Females usually begin egg deposition about three days after attaining maturity. Normally, one egg is deposited about every 36 hours for the first 15 days, but this rate gradually decreases. A female may live up to six weeks and deposit 20-30 eggs. Time for egg-to-adult development averages about one month, but varies with skin temperature of the host and moisture conditions. This species is usually slower to build up to damaging populations than some of the other species found on cattle. The skin of heavily infested animals may become raw and bleeding from constant rubbing in attempts to relieve the itching and irritation caused by the lice. When these raw areas scab over, lesions may somewhat resemble those produced by scab mites (Chapter 10); differential diagnosis is made by identification of the parasite.

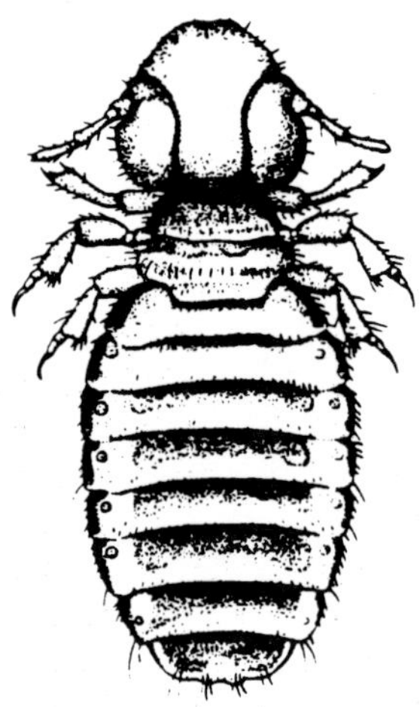

Figure 5.10. Biting cattle louse. (PCEX)

The short-nosed cattle louse, *Haematopinus eurysternus* (Fig. 5.11), is one of the largest louse species found on cattle; it is world-wide in distribution. This species occurs more commonly on mature cattle although younger animals may also be infested. Females are about 3 mm in length; males are slightly smaller. The head is about half as wide as the thorax, short, nearly as broad as long, and bluntly rounded in front. The head and thorax are yellowish-brown and the abdomen is slate blue in color. This species may be readily distinguished from the long-nosed cattle louse by its comparatively short head and by the cone-shaped plates on the lateral margins of the abdomen. The principal breeding site during the winter is the top of the neck; secondary sites are the sides of the neck, the dewlap, back, base of the horns and base of the tail. During the summer, individual lice usually can be found in small numbers along the edges of the ears or the hair around the dorsal portion of the holves or dew claws. Temperature limitations of this species are similar to those of the cattle biting louse. Eggs do not hatch at temperatures below 77^0 F; development normally begins at about $80\text{-}81^0$ F. Eggs are attached to hairs adjacent to the skin but are subjected to temperature fluctuations as normal hair growth gradually moves them away from the skin surface. Young lice remain close to the skin and are protected from severe tempetature fluctuations by the hair coat; skin temperature seldom falls below 86^0 F and usually fluctuates between $90\text{-}97^0$ F from November to February or March, the period of greatest louse activity. There is a gradual increase in numbers beginning in late August or September and populations usually reach a peak during February. The population decreases abruptly after March and persists at a low level during April, May, June and July. Females deposit 1-4 eggs per day for 15 days or longer. Unlike other cattle lice eggs, these are hard-shelled, opaque and brown to brownish-white in color. Hatching occurs in 9-19 days, and nymphs reach maturity in 9-15 days. The life cycle from egg to adult is completed in 20-41 days. Infestations usually remain unnoticed until populations become large enough to cause annoyance. Poor, weak, or otherwise unthrifty animals are the first to show signs of lousiness. Infested animals rub against trees, fence posts and other objects to alleviate itching and irritation resulting in loss of hair, raw skin and bruised areas. Cattle that harbor lice usually do not respond to feed properly and are more prone to disease due to their weakened condition. Anemia may result from blood loss. Young calves may be stunted and, until infestations diminish in late spring or summer, fail to respond fully to improved spring grazing and increased milk production from cows.

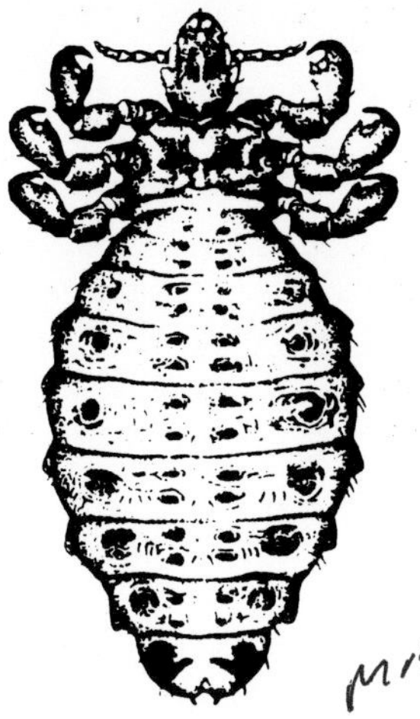

Figure 5.11. Short-nosed cattle louse. (PCEX)

Figure 5.12. Long-nosed cattle louse. (PCEX)

Long-nosed cattle lice, *Linognathus vituli* (Fig. 5.12), often occur on young animals and dairy cattle. Heavy infestations on mature beef animals are more likely to be the short-nosed cattle louse. The long-nosed cattle louse may be distinguished readily from other cattle lice by its longer "nose", or rostrum, and narrow body which gives it a slender, rather than robust, appearance. The abdomen is bluish-black in color and the lateral margins are smooth. This species has been found on cattle throughout the world. The most common sites of infestation are the dewlap and shoulders, although lice may occur on any part of the animal's body, especially the sides of the neck, top line, belly and perianal region, when infestations are severe. As spring approaches and populations begin to decline, the remaining lice seem to migrate toward the shoulder or scrotal areas. Females deposit about one egg per day. The eggs are oval-shaped and blue in color. Hatching occurs in 8-13 days; nymphs require 15-17 days to matute. The complete life cycle requires 23-30 days. The type of damage caused by this species is basically the same as that resulting from short-nosed cattle louse infestation, but the long-nosed cattle louse seldom is as numerous.

The cattle tail louse, *Haemalopinus quadripertusus*, resembles the short-nosed cattle louse with which it is often confused; however, the tail louse is larger (4.5 mm long) and the head, thorax and abdomen are dark brown in color. This tropical species apparently was introduced into Florida, from which it spread along the Gulf Coast to other southern states. Eggs are deposited almost exclusively in the switch of the animal's tail. As eggs hatch, the young nymphs migrate up the tail to the soft-skinned areas of the tail, head and neck. Calves may become heavily infested about the face as a result of sucking the dam through the rear legs and contacting the young nymphs on the switch. The life cycle is not completely known, but the eggs apparently require about 11 days to hatch. This species causes essentially the same damage as does the short-nosed cattle louse. Nymphs and adults seem to prefer tender areas of the skin such as the tailhead, ears and eyelids. Cattle tail louse infestations usually are more difficult to control than those of other species.

The "little blue" or "capillate" louse, *Solenopotes capillatus* (Fig. 5.13), was probably a parasite of deer before it became adapted to cattle since all other members of the genus *Solenopotes* are parasites of the cervid (deer) family. The little blue" louse often is confused with the long-nosed cattle louse, but it is smaller and more robust in appearance. The head is short and broadly rounded instead of long and pointed. The smallest of the sucking lice found on cattle, it is more common throughout the eastern United States than other regions. The "little blue" louse usually clusters on the front part of the animal, especially on the head, dewlap and neck; however, infestations have also been noted along the top line and under the tail. Little is known about the life history except that the eggs hatch in approximately 13 days. Damage is essentially the same as that caused by other species of sucking lice.

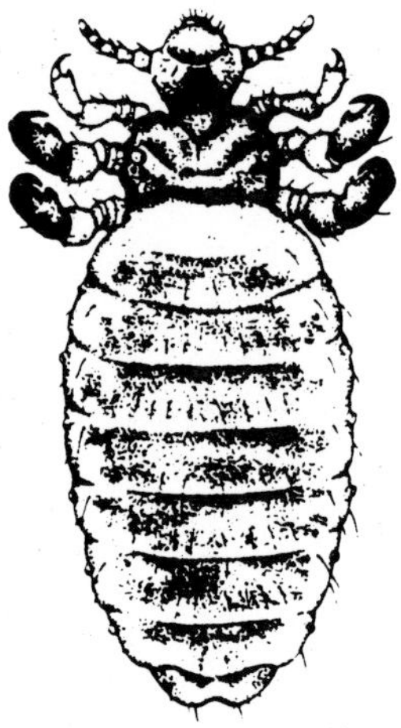

Figure 5.13. "Little blue" or "capillate" louse. (PCEX)

HORSE LICE

Two species of lice are commonly found on horses in the United States; one of these, *Haematopinus asini* (Fig. 5.14), is a sucking louse and the other, *Bovicola equi* (Fig. 5.15), a biting louse. Eggs of the sucking louse are attached firmly to body hairs, usually close to the skin; hatching occurs within 12-14 days. Nymphs mature in 11-12 days and adult females begin depositing eggs. The horse biting louse deposits its eggs similarly to those of the sucking louse; the incubation period is 8-10 days. During cold weather when the hair on horses is long, populations of both species increase rapidly. Preferred areas are the sides of the neck, the flanks and under the jaws; however, lice may be found all over the body and legs of heavily infested horses. When the animals shed their hair in the spring, the lice seem to disappear. Although some individuals usually remain on the host throughout the summer, they are too few to be annoying and are not easily detected. Animals, therefore, usually reenter the winter already infested. The sucking louse feeds by puncturing the skin and sucking blood; the biting louse feeds on particles of hair, scales and skin exudations. Infested horses rub, bite themselves, often kick and stamp, and exhibit extreme uneasiness and annoyance. When infestations are heavy, the hair coat becomes rough and bare patches may appear; rubbing may also result in bruises or skin wounds.

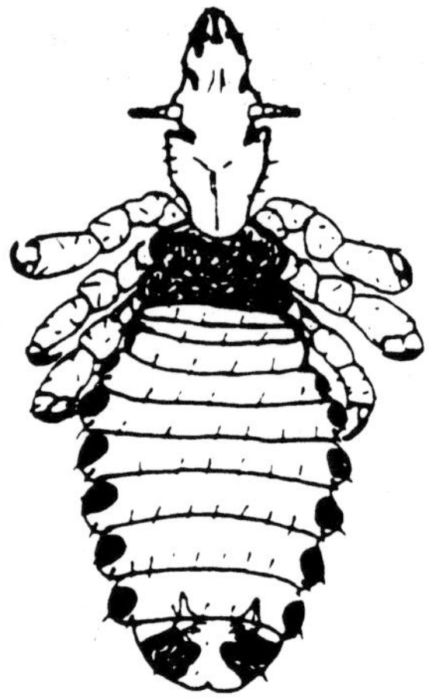

Figure 5.14. Horse sucking louse. (TAEX)

Figure 5.15. Horse biting louse. (TAEX)

HOG LOUSE

Only one species of lice occurs on domestic swine and it is a sucking louse, *Haematopinus suis* (Fig. 5.16). It is one of the largest louse species of domestic animals, 5-6 mm in length; adults are usually bluish-gray in color. Each female louse attaches 1-20 eggs on a single hair bristle and may lay as many as 90 eggs over a 25-day period. Eggs hatch in 12-20 days depending on environmental conditions; young lice attach and begin feeding immediately. Hog lice reach sexual maturity in 13-15 days. These lice prefer to feed on the tender skin behind the ears but they may be found over the entire body. Severe infestations may result in serious weight loss because of poor feed utilization and general unthriftiness.

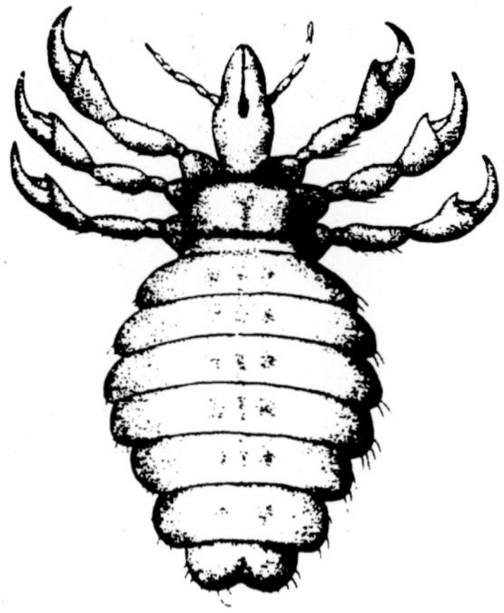

Figure 5.16. Hog louse. (PCEX)

SHEEP LICE

Three species of lice are known to parasitize sheep in the United States. Two of these are sucking lice and the other is a biting or chewing species. The sheep biting louse, *Bovicola ovis*(Fig. 5.17), is one of the most common lice found on sheep; it is distributed widely over the entire United States. It is a small species, about 1 mm in length, mostly white in color although the head and thorax are somewhat darker. The head is about as wide as long with a broad, rounded front. Infestations occur over all areas of the body but the upper sides of the animal are apparently favored. This species moves rapidly over the wool fibers but is usually found near the skin. The female deposits about two eggs, attached to the wool by a viscid substance, every three days. The eggs hatch in 9-10 days, and the nymphs mature in about 21 days. Biting lice live on scurf and other skin and wool fragments. When feeding close to the skin, they cause sheep to bite and pull their wool and to rub against posts, trees and other objects. The fleece of heavily infested sheep becomes ragged and torn and is greatly reduced in value.

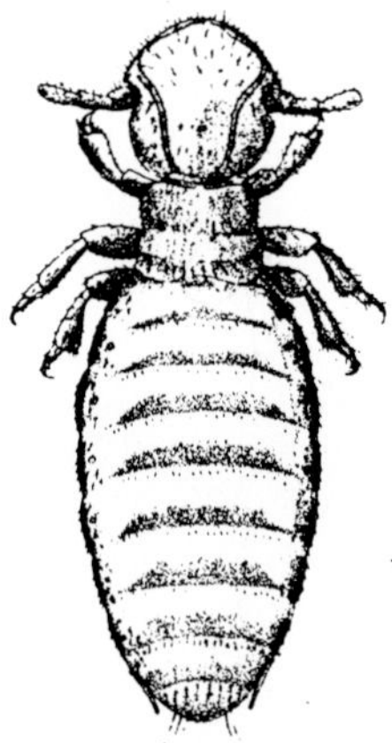

Figure 5.17. Sheep biting louse. (PCEX)

Records of occurrence of the sucking body louse, *Linognathus ovillus* (Fig. 5.18), in the United States are somewhat confusing and the distribution pattern is not completely known. The louse occurs commonly in colonies on various parts of the body including the face. The body is yellowish-brown in color, and the head is somewhat longer than the thorax. Males and females are about 2 and 2.5 mm in length, respectively. Usually, the wool in the infested area is discolored and contains numerous brown particles, the excretions of the lice. All stages of the sucking body louse are found on the host; however, individuals may live 3-7 days off the host without food, and eggs may hatch several weeks after being separated from the host. Eggs are glued to the hair and hatch in 10-18 days. Development may be completed and new eggs deposited within two weeks of hatching. Irritation interferes with feeding causing decreased weight gains; scratching results in wool loss, cuts and bruises. Excretions of the lice soil the wool.

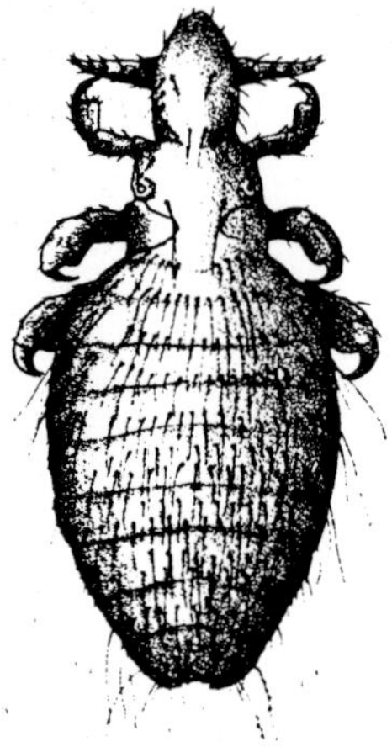

Fig. 5.18. Sheep sucking body louse. (PCEX)

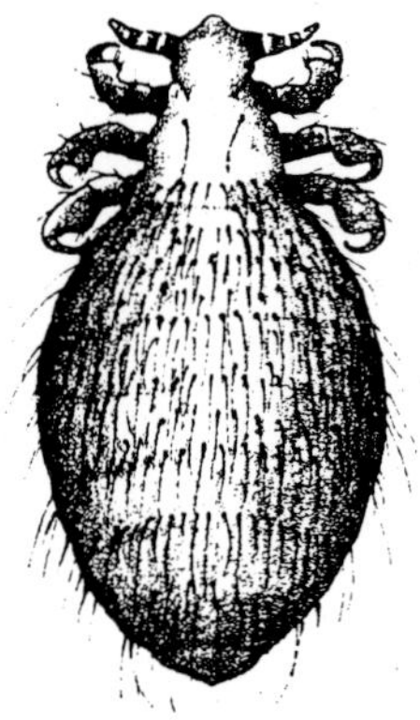

Figure 5.19. Sheep foot louse. (PCEX)

The sucking foot louse, *Linognathus pedalis* (Fig. 5.19), is similar in appearance to the sucking body louse. The female is about 2 mm long and 1 mm wide; the male is broader and flatter. The abdomen is thickly covered with long, slender bristles. It is distibuted throughout the United States wherever sheep are produced. Occasionally, goats also become infested by this species. The foot louse infests the lower, hairy parts of the body including the shanks, belly and scrotum. It is most active from November to June; populations decrease during the warmer, summer months and the lice become inconspicuous. Young lambs are more heavily infested than older sheep and can be found harboring lice within 48 hours of birth. Eggs are deposited at the rate of one per day. The complete life cycle averages 40-45 days. Damage caused by the foot louse is similar to that caused by the sucking body louse except that the foot louse is considerably less harmful. Systemic effects are somewhat less severe and wool damage is less important since feeding occurs on the hairier parts of the animal's body. Severe infestations may cause slight lameness.

GOAT LICE

Several species of lice infest goats and different species usually occur on Spanish goats and Angora goats, although some intermingling of species may result when both types of goats occur together. Two species of blood-sucking lice, the goat sucking louse, *Linognathus stenopsis* (Fig 5.20), and the closely related species, *Linognathus africanus*, parasitize both types of goats. These species are similar in appearance and are distributed widely throughout the United States. Three species of chewing or biting lice also occur on goats. *Bovicola crassipes* (Fig. 5.21), a large, yellow louse, and *Bovicola limbata*, the Angora goat biting louse Fig. 5.22), are normally found on Angora goats; the goat biting louse, *Bovicola caprae*, is commonly found on Spanish goats. These species are found throughout the United States. All of the lice attacking goats spend their entire life upon the host unless dislodged.

They may, however, transfer from one host to another when animals are in close contact. The eggs are attached to the hairs by a viscid substance produced by the female. Eggs hatch in 5-18 days and the young lice mature 14-21 days later. The usual symptoms of scratching and rubbing depend upon the number of lice present. The blood-sucking species are more injurious than the biting lice because they may cause the formation of scabby, bleeding areas. In Angora goats, the clip of mohair is sometimes reduced in value as much as 10-25% by biting lice.

Figure 5.20. Goat sucking louse. (TAEX)

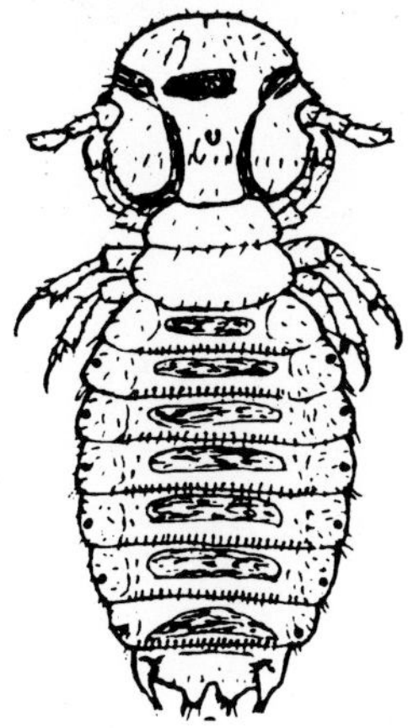

Figure 5.21. Goat biting louse. (TAEX)

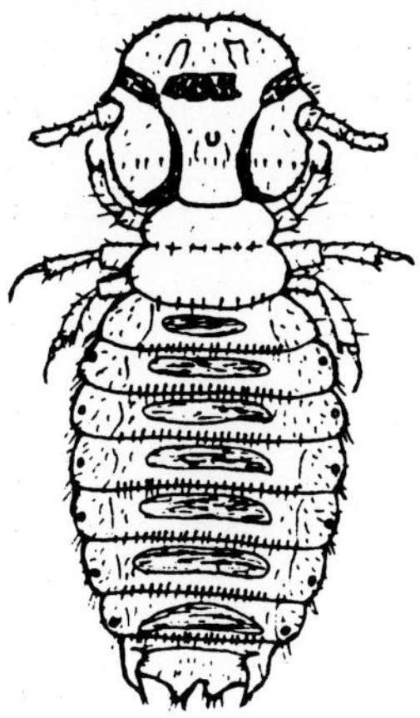

Figure 5.22. Angora goat biting louse. (TAEX)

PET ANIMAL LICE

Although rarely found, the dog biting louse, *Trichodectes canis* (Fig. 5.23), and the cat biting louse, *Felicola subrostratus* (Fig. 5.24), may occur as parasites of dogs and cats, respectively. Both species may produce considerable irritation and hair loss may result from severe infestations. The dog biting louse may also serve as an intermediate host of the double-pored tapeworm of dogs, *Dipylidium caninum.*

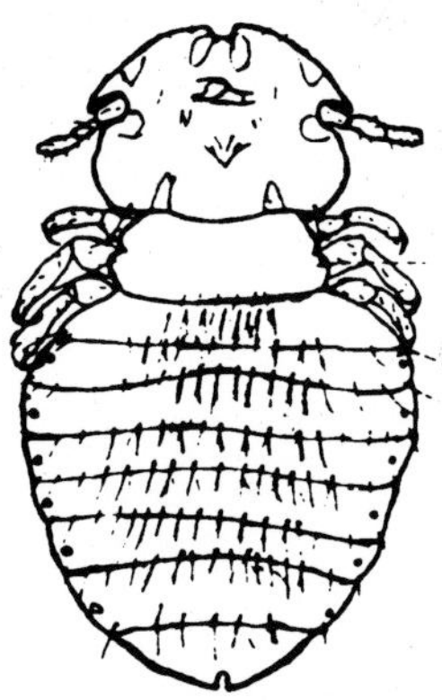

Figure 5.23 Dog biting louse. (TAEX)

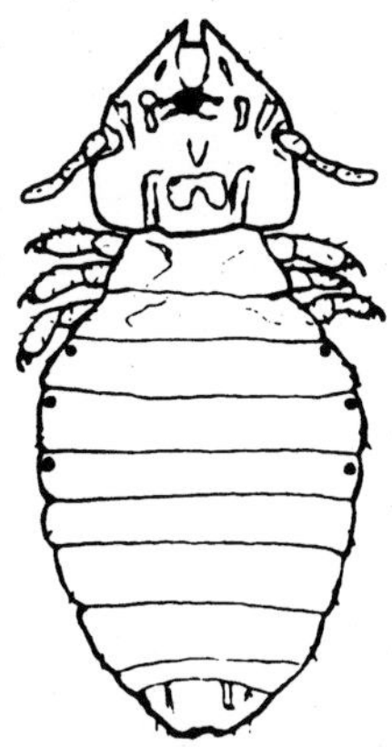

Figure 5.24. Cat biting louse. (TAEX)

Pet birds may be attacked by several species of biting lice. Symptoms and nature of damage are essentially similar to those produced in domestic poultry.

LICE CONTROL

Spraying or dusting of louse infested poultry with approved insecticides is effective for immediate and complete control of these parasites. Chemical formulations are available as wettable powders or emulsifiable concentrates for application directly to infested poultry; insecticidal dusts may be applied to litter and nest boxes as well. Sprays should be applied to caged birds from underneath with sufficient pressure (40 psi) to penetrate the feathers to the skin surface of the host.

The most common manner of treating body lice of livestock is by spraying during the fall, winter and spring months. Power sprayers capable of delivering 250-400 psi of pressure are necessary to wet infested animals to their skin. Since horses do not tolerate spraying well, they may be individually treated by briskly and thoroughly brushing the insecticide onto louse infested areas. It is usually necessary to retreat all infested animals approximately two weeks later in order to kill young nymphs that have hatched from eggs not affected by the initial treatment. Systemic pour-on and spot-on insecticides are convenient to apply and do not stress animals by wetting them during cold weather; they are usually formulated as ready-to-use solutions or as liquids to mix with water and are applied to the backline of the animal where they are absorbed through the skin and dispersed throughout the body. Systemic treatements are generally more effective against sucking lice than biting lice. Caution should be used in applying systemic insecticides to cattle not previously treated for grubs due to possible adverse side effects. Insecticidal dust bags placed in strategic sites (e.g., near water or feed troughs, entrances to loafing areas, etc.) also aid in maintaining louse populations at low levels on cattle.

CHAPTER 6
FLEAS

Fleas are small (2-4 mm), wingless parasites of mammals and birds. The name of the order, "Siphonaptera", refers to their method of feeding through a siphon or tube and their absence of wings. Since the adults are laterally compressed and possess posteriorly directed spines, they are well adapted for movement through the hairs and feathers of their hosts. The mesothoracic legs are superbly adapted for jumping; some species may leap as high as 7-8 inches vertically and 14-16 inches horizontally. Most species move about a great deal and remain on the host only part of the time. Mouthparts of the adult consist of three stylets (Fig 2.5). The paired maxillae, acting as cutting organs, along with the epipharynx penetrate the skin of the host. When a small blood vessel is encountered, the tip of the epipharynx enters the lumen while the maxillae, which form the salivary canal, remain outside periodically emitting salivary secrecions. Both sexes are blood feeders and females require a blood meal to produce eggs. Most fleas feed at frequent intervals; they are easily disturbed and seldom complete a meal at one feeding.

Most species of fleas are obligatory ectoparasites of smaller mammals such as rodents, rabbits, moles and bats although several are parasitic upon larger animals and birds. Some fleas are quite specific in their host preference, feeding on only one type of host (i.e., closely related species of the same genus) while others have developed the ability to feed upon numerous host species. Fleas are very sensitive to extremes of temperature and humidity which explains the relative abundance of fleas that infest mammals and birds living in burrows and sheltered nests and the light infestations of hosts that either have no permanent abode or live in nests exposed to the elements. Since nests furnish an abundance of organic food for flea larvae, the rate of survival is high. Fleas infesting burrowing or nocturnal animals tend to have poorly developed eyes or eyes may be absent. Those infesting animals active during the day are more likely to have well-developed eyes. Researchers have found that the breeding cycle of the European rabbit flea is regulated by the reproductive hormones of the pregnant rabbit. Reproduction of other species such as the Oriental rat flea, cat flea and certain bird fleas, however, is not dependent on the hormonal cycle of the host although flea populations are often greatest during the host's breeding season.

GENERALIZED LIFE CYCLE

Fleas undergo complete metamorphosis and pass through four stages of development: egg, larva, pupa and adult (Fig. 6.1). The time for completion of the life cycle varies according to species, temperature, humidity and food. Under favorable conditions, a generation can be completed in as little as 2-3 weeks although the average is approximately 35 days; several months or more may be required during cold weather.

Mating usually occurs on the host. The sticktight flea, *Echidnophaga gallinacea*, and the chigoe, *Tunga penetrans*, are sessile; the female remains attached in one location for long periods of time and mating occurs while she is in the feeding position. Eggs are usually deposited among the hairs or feathers of the host or in the nest. They are smooth, spherical to oval in shape, light in color, and large enough to be seen with the naked eye. Since they are not sticky, the eggs drop to the ground or into the nest or bedding resulting in the subsequent high concentrations of adult fleas in dog or cat premises and in areas of buildings frequently used by the host. Females do not lay their full quota of eggs at one time, but after repeated blood meals which are necessary for ovarian development. Successive matings are not necessary for fertilization as sperm from the initial mating are stored in the spermatheca of the female and used as required. Eggs hatch in two days to several weeks depending upon temperature and humidity.

Flea larvae are small, 13-segmented, worm-like organisms lacking legs; mouthparts are adapted for chewing. They are found most frequently in houses in floor cracks and in rugs, as well as in kennels, stables, chicken coops, animal burrows and nests. Larvae feed on all types of organic detritus as well as feces of the adult flea which contain partially digested blood. The three larval stages usually require approximately two weeks to complete development.

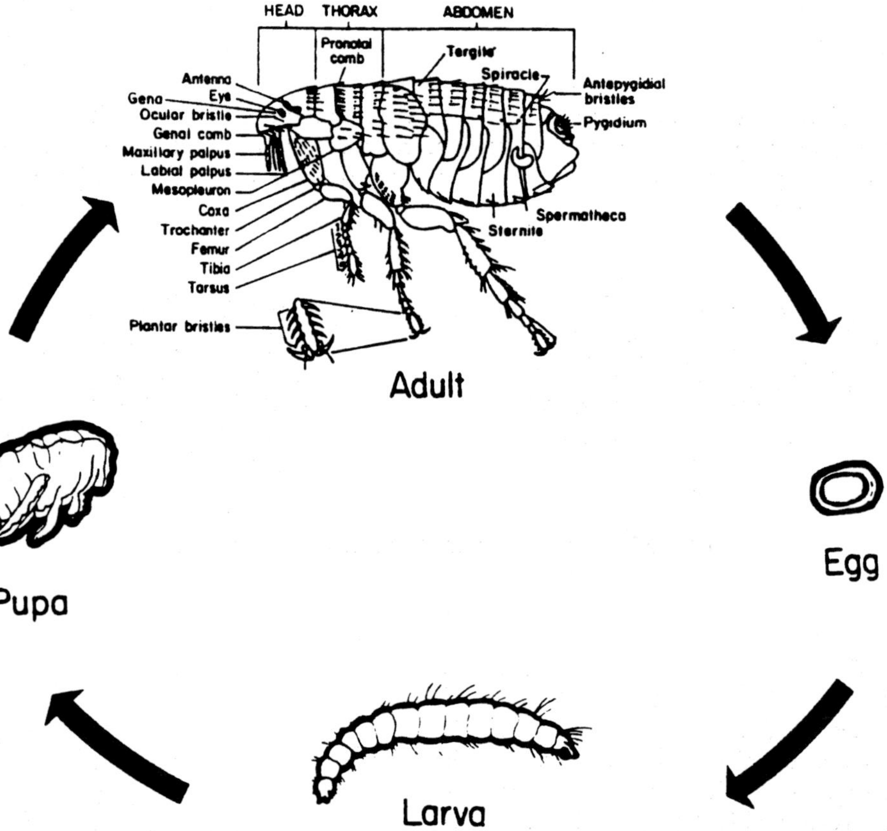

Figure 6.1. Flea life cycle. (CDC)

Pupae are usually enclosed in cocoons of finely spun silk encrusted with granules of sand or various types of organic debris. The pupal stage is usually completed within about two weeks, but the newly formed adult may remain within the cocoon until it emerges in response to a stimulus that varies with the species. Cat fleas emerge in response either to vibrations or the emission of carbon dioxide from a nearby host. Certain wild rodent fleas emerge in response to an increase in humidity which may occur only once a year in some desert areas.

Adults are usually ready to feed within 24 hours of emergence from the cocoon. Mating usually follows the initial blood meal. Very little is known about the time needed to complete the life cycle of most species or how many broods may be produced in a year. Some species, such as cat and dog fleas, oriental rat fleas and northern rat fleas, apparently breed continuously, a generation taking a few weeks to a month or more in warm weather and longer during the cooler portion of the year. Other species, such as many bird fleas in the genus *Ceratophyllus*, have only a single generation per year. Many of the "nest fleas" of North Ametican rodents are far more abundant during the winter.

SOME COMMON SPECIES OF FLEAS

Cat and Dog Fleas: *Ctenocephalides felis* and *Ctenocephalides canis*

Cat and dog fleas, *Ctenocephalides felis* and *Ctenocephalides canis* (Fig. 6.2), respectively, are world-wide in distribution. Both species occur throughout the United States, however, they are less common in the Rocky Mountain area. Cat fleas appear to be more abundant and more generally distributed than dog fleas. The head is about twice as long as high in the cat flea while it is only about as long as high in the dog flea. Moreover, the front margin of the head is more acute in the cat flea than in the dog flea (Fig. 6.2). In most cat fleas the first and second teeth of the genal comb are approximately equal in length, while the first tooth is shorter than the second in dog fleas. Under favorable conditions, a generation requires from 2-4 days for the eggs, 8-24 days for the larvae and 7-10 days for the pupae. These fleas are commonly found in homes, under houses or in yards where pets (dogs and cats) are kept but prefer locations where dust and organic debris accumulate. Both species may also attack dairy calves and young foals maintained on straw or other bedding material in barns and stables. Neither species are known to transmit any diseases to companion animals or livestock, but may serve as intermediate hosts of *Dipylidium caninum*, the double-pored tapeworm of dogs. Severe allergic responses to salivary secretions injected during feeding have been reported in dogs.

Figure 6.2. Head of dog (left) and cat flea (right). (CDC)

Human Flea: *Pulex irritans*

Human fleas (Fig. 3) are distributed throughout the South and Middle West where they may be pests of swine and associated premises; infestations may persist for weeks or months after premises have been vacated. The human flea also attacks a wide variety of other hosts including dogs, coyotes, prairie dogs, ground squirrels and burrowing owls. Human fleas can be distinguished from other common fleas by the absence of pronotal and genal combs, the ocular bristle being inserted beneath the eye and the absence of an internal, rod-like thickening on the mesopleuron (Fig. 6.3). A related species, *Pulex simulans*, occurs in the central and southwestern United States and in Central and South America.

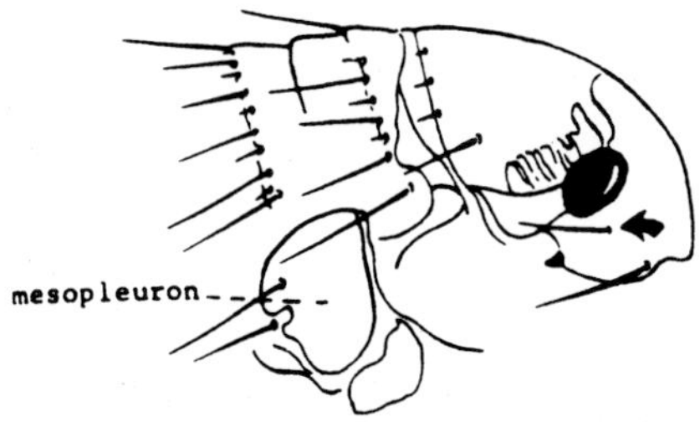

Figure 6.3. Head of human flea. (CDC)

Oriental Rat Flea: *Xenopsylla cheopis*

The Oriental rat flea is the primary vector of bubonic plague throughout much of the world. This species was first collected in the Nile Valley; hence the species name “cheopis” for Cheops, the Pharaoh who constructed the Great Pyramid at Giza. The Oriental rat flea has been introduced into all sections of the world with Norway and roof rats. The species is established throughout most of the United States, being one of the most abundant rat fleas in the South; it is found as far north as New Hampshire, Minnesota and Washington where it is abundant during the summer and fall, becoming scarce during the winter. The life cycle varies, but may be completed in as few as 4-8 weeks. Adults lack both genal and pronotal combs. the ocular bristle is located in front of the eye and the mesopleuron has a vertical rod-like thickening (Fig. 6.4). *Xenopsylla cheopis* is the only species in the United States whose female has a dark-colored spermatheca.

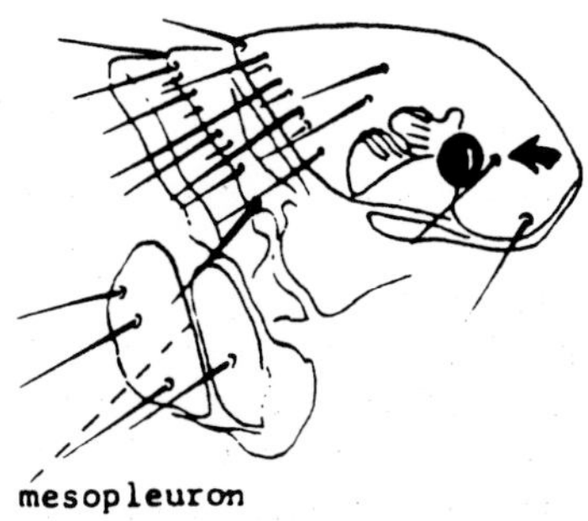

Figure 6.4. Head of Oriental rat flea. (CDC)

Sticktight or Southern Chicken Flea: *Echidnophaga gallinacea*

Sticktight fleas (Fig. 6.5) are found in the southern United States from South Carolina to California. Adults are characterized by the absence of both genal and pronotal combs, and a distinctive angular margin to the front of the head. This species attacks cats, dogs, horses and humans; however, it is most commonly associated with chickens and turkeys maintained on litter in old, poorly managed houses. It is not a common parasite in modern commercial poultry operations. Adults attach in clusters to the skin around the eyes, comb and wattles. Heavy infestations cause intense irritation, reduced egg production, blood loss and ulcers in areas where females embed their heads in the host's skin. Ulcers around the eye may result in blindness and eventually death due to an inability of the host to feed.

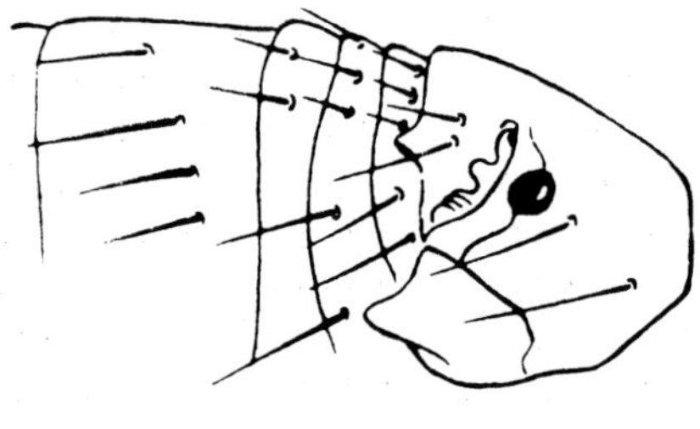

Figure 6.5. Head of sticktight flea. (CDC)

Chigoe: *Tunga penetrans*

The chigoe flea, or "jigger", is the pest which inspired the sailor's oath "I'll be jiggered". It is a small, burrowing flea found in tropical and subtropical regions in North and South America, the West Indies and Africa; it is not known to be well established in the United States. This species is unusual in that the female actually burrows into or becomes embedded in the skin of the host. Engorgement with blood and the development of eggs cause great distention of the flea abdomen resulting in considerable tissue inflammation and, possibly, the formation of uclerative lesions on the host.

Rodent Fleas

Numerous species of fleas (e.g.. the northern rat flea, *Nosopsyllus fasciatus*; mouse flea, *Leptosylla segnis*; rabbit flea, *Cediopsylla simplex*; squirrel flea. *Orchopeas howardii* and ground squirrel flea, *Diamanus montanus*) commonly parasitize a wide variety of rodents and other small mammals throughout the United States; they also occur on dogs and cats, but only rarely. These species are important primarily as vectors of rural plague among the various host species.

FLEA CONTROL

On Pets

Simultaneous treatment of both hosts and premises is necessary for effective flea control, and reinfestation will occur within a short period of time unless control procedures are thorough. Thus, a cat or dog treated to control fleas will soon become infested again unless all nearby flea breeding sites are also treated. It is necessary to be familiar with the habits of the host so as to learn its favorite resting places and to treat them. Treatment of the infested animal plus complete coverage of all breeding areas is much more effective than periodic treatments of the host alone.

Approved insecticides may be applied to a host as a spray, dust or bath (dip). Applications should be made evenly over the entire body of the animal; care should be taken to avoid getting the insecticide into the eyes, nostrils or mouth of the host. Retreat as needed and always follow label directions. Resin collars impregnated with insecticides are approved for use on dogs, except whippets and greyhounds, and cats, except Persian; they usually provide some degree of protection for 2-3 months. Flea collars should be used with care since individual animals may be sensitive to the insecticide. A relatively new development in the control of fleas and other ectoparasites on dogs is the administration of one of the less toxic systemic insecticides; pills containing the insecticide are sold and fed to the animal only by, or on the order of, a licensed veterinarian.

When treating premises, special attention should be given to the resting places of the host as flea eggs, larvae, pupae and adults are most abundant at these sites. Infestations may be greatest in bedding used by the pets, in rugs, under porches and similar areas. When possible, the animal's bedding should be laundered in hot, soapy water or burned. A vacuum cleaner may be used to remove accumulations of lint and dust that contain flea larvae and pupae. Infested premises may then be treated with a residual insecticide. A number of factors (e.g., odor, fire hazard, possibility of staining and use of the treated area) will influence the choice of insecticide and formulation for indoor applications. Dusts may also be used to control fleas inside buildings, particularly in animal boxes or bedding, basements and other areas where the whitish discoloration from powders is not objectionable. Dusts are frequently blown under porches and into crawl spaces under homes, garages and outbuildings.

In many places in the United States, particularly in the South and West, yard infestations often become a serious problem. For insecticidal treatment of lawns and yards, dusts or suspensions are frequently used in preference to emulsions or solutions. Wettable powders and dusts are safe for use on shrubbery and grasses and have long residual actions; however, they leave a light-colored deposit. Some of the auxiliary solvents used in preparing emulsifiable concentrates may "burn" foliage or grass. The use of oil solutions on vegetation is even more dangerous since many of the petroleum solvents are known to be phytotoxic. Soil or lawn areas that require treatment should be very thoroughly soaked.

On Livestock and Poultry

Initial control should be aimed at killing adult fleas on animals and poultry with secondary attention given to either removal of infested animal bedding or thorough treatment of these materials. Flea-infested livestock should be removed from their quarters and thoroughly treated in small groups or individually hand-brushed with insecticide solutions. Before their return and after removal of old animal bedding, an insecticide should also be applied to the floor and wall cracks of the premises. As a preventive measure, animal housing should be rodent-proofed and pets should be denied access to animal quarters. Flea-infested poultry can be sprayed or dusted in small groups and insecticidal dusts may also be placed in dust boxes or in soil wallows in houses with large flocks. Dust formulations placed in nests should control fleas on pigeons and other nesting birds.

PLAGUE

Two main epidemiological types of plague have been described (Fig. 6.6). Demic plague (from *demos* = people) refers to primary pneumonic plague involving man-to-man transmission by airborne infection, or, rarely, man-to-man bubonic plague transmitted by fleas. Zootic plague (from *zoo* = animal) refers to transmission from lower animals, especially rodents, to man through the bites of infected fleas or by contact with infected animals. This latter type is further divided into: (a) urban plague, the classical form of the disease, contacted in cities where people have close contact with domestic rats (*Rattus*) and their fleas (particularly *Xenopsylla*) and (b) rural plague, also known as sylvatic or campestral plague, acquired in rural or wilderness areas after contact with wild rodents, lagomorphs or their fleas. Rural plague is endemic throughout much of the southwestern United States.

Fleas becomes infected after feeding on the blood of an infected host. The bacteria, *Yersinia pestis*, multiplies within the proventriculus and stomach of the flea, resulting in the formation of an obstruction; such a flea is referred to as "blocked". When a "blocked" flea subsequently attempts to feed on man or some other animal, the blood, which cannot pass beyond the blockage, becomes contaminated with plague bacilli. When this contaminated blood is regurgitated into the bite injury, plague organisms gain entry into the bitten animal. "Blocked fleas" live for variable lengths of time during which they remain hungry and make repeated efforts to feed; the tendency for a flea to become blocked varies with the species and may be taken as a measure of its vector efficiency. The Oriental rat flea is the most important vector of urban plague because of the ease with which it becomes blocked, its tendency to feed on both infected rodents and man, and its abundance near human habitations. Various species of wild rodent fleas serve as vectors of rural plague in the United States.

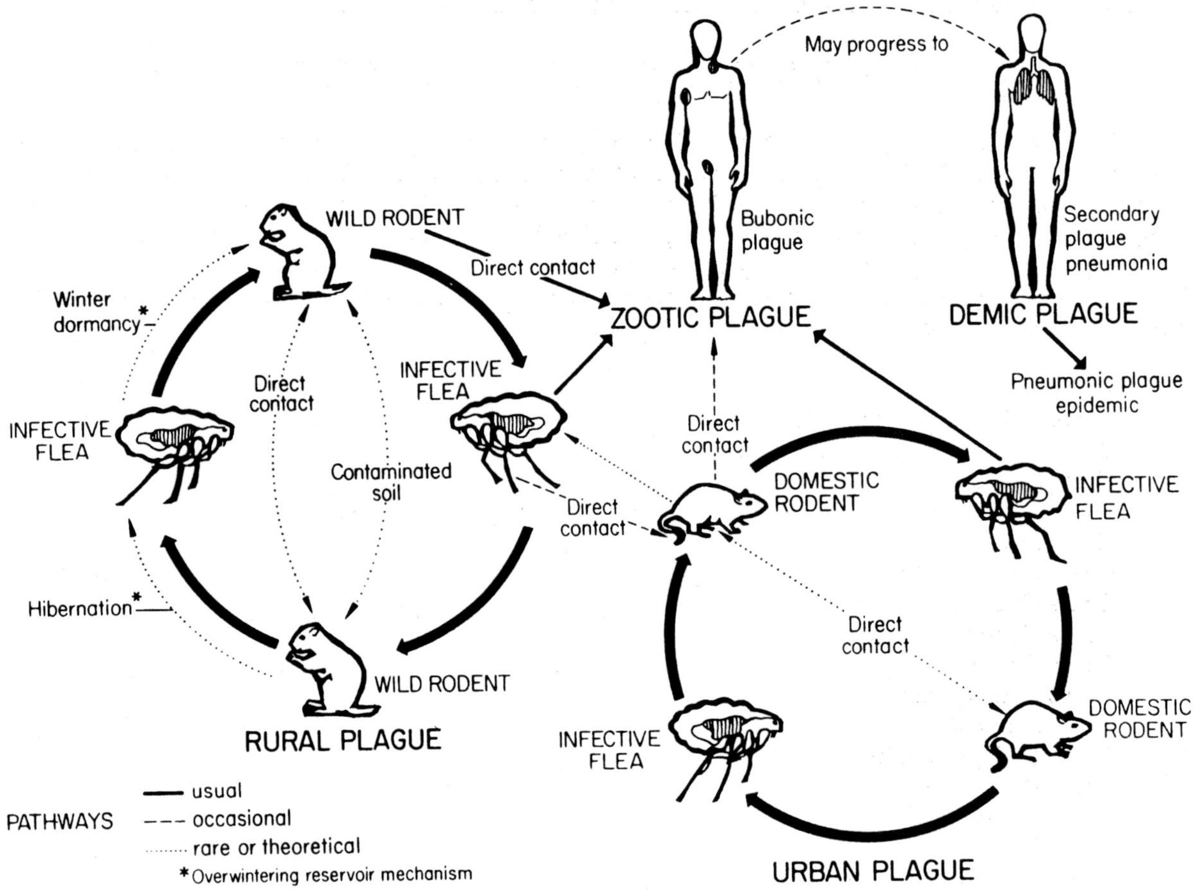

Figure 6.6. Epidemiological cycles of plague. (CDC)

CHAPTER 7

DIPTERA: MOSQUITOES AND BITING GNATS

MOSQUITOES (FAMILY CULICIDAE)

Mosquitoes, belonging to the family Culicidae, are the most prominent of the numerous arthropods attacking man and animals. They are the sole vectors of malaria, yellow fever and dengue to man and are important in the transmission of dog heartworm and viral encephalitides to horses. Adults differ from other species of flies in having three characters in combination: (1) an elongate proboscis, (2) long, many-segmented antennae and (3) scales on the wing veins and wing margins. Mosquitoes constitute a very large group containing over 2.500 species; over 150 species belonging to 13 genera distributed among three subfamilies exist in temperate North America. Their general classification is outlined below:

Order Diptera (flies)
- **Family Culicidae (mosquitoes)**
- **Subfamily Anophelinae (anophelines)**
- Genus *Anopheles*
- **Subfamily Culicinae (culicines)**
- Genus *Aedes*
- Genus *Coquillettidia*
- Genus *Culex*
- Genus *Culiseta*
- Genus *Deinocerites*
- Genus *Haemagogus*
- Genus *Mansonia*
- Genus *Orthopodomyia*
- Genus *Psorophora*
- Genus *Uranotaenia*
- Genus *Wyeomyia*
- **Subfamily Toxorhynchitinae**
- Genus *Toxorhynchites*

GENERALIZED LIFE CYCLE

The life cycle of mosquitoes is divided into four stages: egg, larva, pupa and adult (Fig. 7.1). The first three stages are aquatic but the adult is an active, flying insect that ranges far from the aquatic habitat and feeds upon the blood of man and animals or upon plant juices. Eggs are deposited singly or in groups directly on the water surface, or singly in depressions or containers that may subsequently hold water (e.g., soil, tree holes, buckets, tin cans, old tires, etc.). *Anopheles* eggs are laid singly on the water surface, and are provided with a pair of lateral floats (Fig. 7.1). *Culex* and *Culiseta* eggs are deposited side by side to form a raft (Fig. 7.1) that often contains 100 or more eggs. Some species of *Aedes* and all *Psorophora* oviposit on soil and their eggs remain unhatched for weeks or months. After the eggs are flooded, hatching usually occurs within 2-3 days.

Mosquito larvae have three distinct body regions: head, thorax and abdomen (Fig. 7.2). The larval period consists of four developmental stages which collectively require from four days to two weeks to complete depending on such factors as water temperature and food supply. Larvae are aquatic and have adapted to almost all kinds of habitats except flowing streams and the open waters of large streams, lakes and seas. Some species occur in permanent ponds and marshes, others in temporary flood waters of woodland pools and still others in water contained in tree holes, leaves of plants or artificial containers. Although mosquito larvae obtain food from the water in which they live, they must come to the surface to breathe or, as in the case of *Mansonia and Coquillettidia*, obtain air from the underwater portions of plants. The air tube of culicines and the spiracular plate and palmate hairs of anophelines

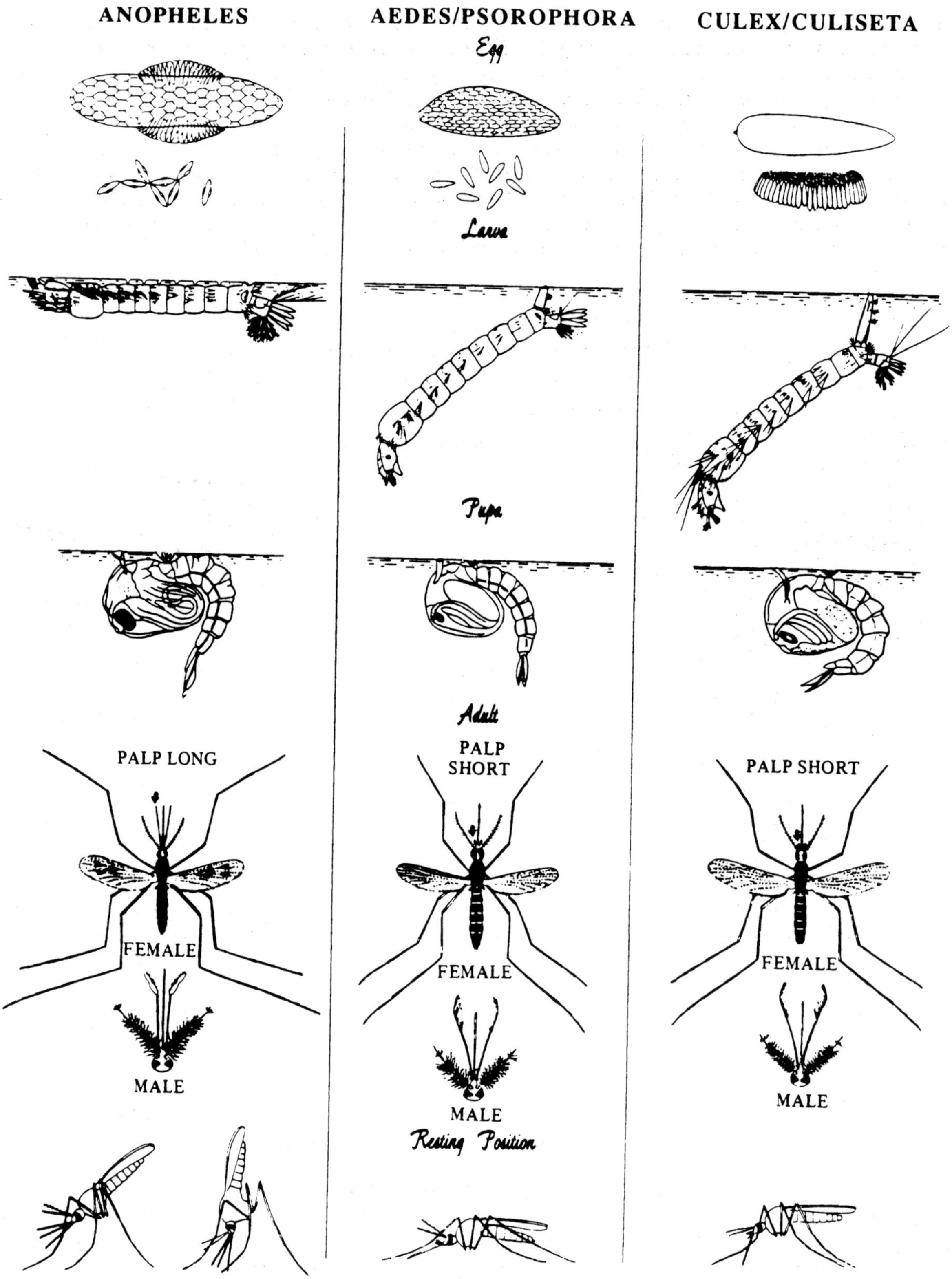

Figure 7.1. Life cycles and characteristics of anopheline and culicine mosquitoes. (CDC)

suspend the larvae from the water surface film. Anopheline larvae lie parallel to the water surface (Fig. 7.1), while most other groups hang head down with only the tip of the air tube penetrating the surface film (Fig. 7.1). Mouthparts consist of a series of brushes in addition to grinding and grasping structures allowing them to strain out small aquatic organisms and particles of plant material present in the water, a few predaceous species have mouthparts adapted for grasping prey. Locomotion is accomplished through jerks of the body and by propulsion with the mouth brushes. Movement of anopheline larvae at the surface is generally of the first type. "Crawling" movements of culicine larvae along the bottom and their slow movements at the water surface probably result from propulsive action by the mouth brushes. Mosquito larvae are affected by light and water conditions including temperature, movement, dissolved gases and salts, and other living organisms present in the habitat. Aquatic vegetation is important for protection from predators such as fish and insects which destroy large numbers of mosquito larvae.

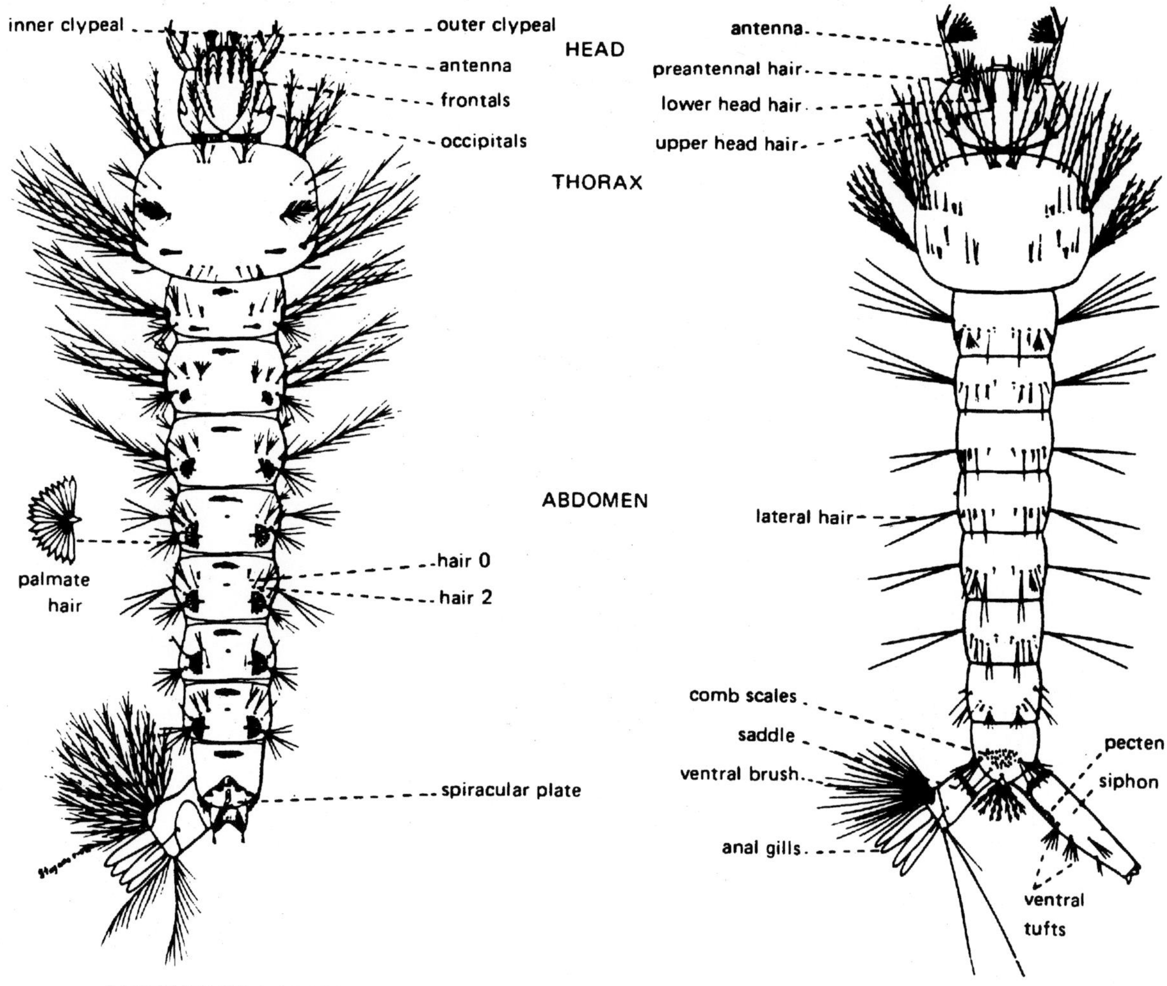

Fiure 7.2. Fourth stage *Anopheles* and *Culex* larvae. (CDC)

Like the larvae, mosquito pupa are also aquatic. Although this stage does not feed, pupae are very active and, except for *Mansonia* and *Coquillettidia*, must come to the surface for air. Most species are lighter than water, their buoyancy being due to an air space between the wing cases on the underside of the combined head and thorax. By vigorous movements of the abdomen, pupae move sideways or downward with considerable speed, rising directly to the surface when movement stops. Pupal development requires from one day to several weeks, after which the pupal skin splits and the adult emerges onto the surface of the water.

The adult mosquito (Fig. 7.3) is a small, fragile insect with a slender abdomen, one pair of narrow wings and three pairs of long, slender legs. Species vary in length from 2-15 mm. The three body regions (head, thorax and abdomen) are well-defined The head is almost spherical and bears a pair of large compound eyes, a pair of long, filamentous antennae, a pair of palps and a proboscis; it is joined to the thorax by a narrow membranous connection. Antennal segments are ringed by a whorl of hairs which are short and sparse in females but long and bushy in males (Fig. 7.1). The palpi of anopheline females are about as long as the proboscis, while those of the male are enlarged at the tip (Fig. 7.1); in culicines, the palpi of the females are short, while those of the male are usually long, densely haired and pointed (Fig. 7.1). The proboscis projects downward and forward from the lower front margin of the head; it consists of a labium, or sheath-like structure, that encloses a group of six stylets (Fig. 2.5). Mouthparts of the male are incapable of piercing the skin of humans or animals. The thorax, or middle region of the body, bears the wings and legs as well as coarse hairs and scales that are useful in identification. The elongate, nearly cylindrical abdomen consists of ten segments, only eight of which are readily visible. The ninth and tenth segments are greatly modified for reproductive functions. In *Aedes* and *Psorophora*, the female abdomen is tapered apically, with the eighth segment withdrawn into the seventh; in other U.S. genera the abdomen is bluntly rounded at the apex. The terminal segments of the male abdomen are greatly modified for mating and are of value in identification of the species.

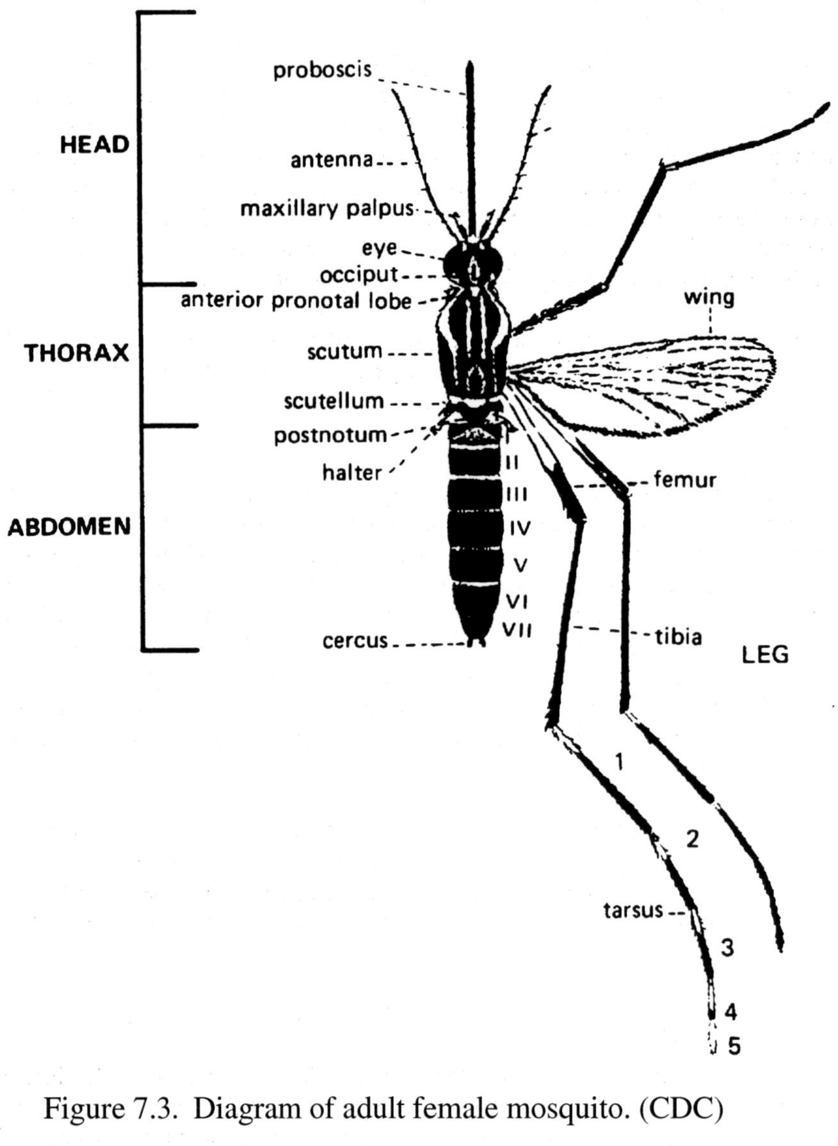

Figure 7.3. Diagram of adult female mosquito. (CDC)

About equal numbers of male and female mosquitoes are produced. Males ordinarily emerge first, remaining near the breeding site, and mate with the females soon after the latter emerge. Only the females suck blood and most, but not all, species require a blood meal for ovarian development. Mosquitoes vary considerably in host preference; some species prefer to feed on birds or on domestic animals such as cattle or horses, while others seem to prefer man. Because these preferences are not exclusive, a species will often feed on alternate hosts, facilitating the transmission of certain disease organisms. Females require two or more days to digest a blood meal, lay a batch of eggs and then seek another blood meal. This cycle of feeding, ovipositing and feeding again may be repeated many times during the life of the female; a single mating is generally sufficient to fertilize her lifetime egg production. The life span of adult mosquitoes is not well known. Some species may live one or two months during the summer, although they may not survive this long under unfavorable conditions. Overwintering adults may live six months or more.

ECOLOGIC AND BIONOMIC TYPES OF MOSQUITO LIFE HISTORIES

For practical purposes, the numerous species of mosquitoes can be grouped on the basis of similarities in their larval habitats, which generally reflect other important differences in their bionomics and ecology (e.g., oviposition habits. developmental patterns, brood patterns, seasonal density and dispersal) as well. A convenient grouping includes: (1) the permanent pool group, (2) the transient water group, (3) the floodwater group and (4) the artificial container and tree-hole group.

Species belonging to the permanent pool group generally occur in fresh bodies of quiet water that is exposed to sunlight and contains an abundance of surface vegetation and flotage at the air-water-plant interface. Typical habitats are shallow margins of ponds, lakes and smaller impoundments whose main characteristic are a degree of permanency. Such habitats are particularly suitable for *Anopheles* and some species of *Culex*, *Coquillettidia* and *Monsonia*. Mosquito species which inhabit permanent waters deposit their eggs on the water surface, singly in the case of *Anopheles*, or in rafts in the case of *Culex*, *Coquilleltidia* and *Mansonia*. Such sites are sheltered from wave action and always contain vegetation that is suitable for larval attachment or protection. Breeding is generally continuous in these species and several generations per year are produced.

The transient water group is composed mostly of *Culiseta* species, some *Culex* and occasionally *Anopheles*. Their habits are similar to those of the permanent pool group, but they seem to be more often associated with temporary pools. Typical sites are roadside ditches, borrow pits, canals, ground pools, clogged streams and irrigated lands.

The floodwater group of mosquitoes includes species of *Aedes* and *Psorophora* which characteristically deposit their eggs singly on damp soil in depressions in grassy areas or woodlands or along vegetated shorelines that are intermittently inundated with water. Salt-marsh mosquitoes are also included in this group. When flooded after a period of desiccation, eggs hatch if conditions (temperature, ph, oxygen concencration, etc.) are favorable; otherwise, eggs may remain dormant and viable on the soil until conditions are more suitable. Typically, large broods are produced in a short period of time. Larval development is uniform and adults may appear as early as six days after flooding. Some species, particularly *Aedes* in northern areas, produce only a single brood per year, while many other species of *Aedes* and *Psorophora* produce multiple broods yearly. Mosquitoes in this group are particularly troublesome as pests and some characteristically fly long distances (5-20 miles) from breeding sites. Woodland species tend to remain near the breeding area.

The artificial container and tree-hole group of mosquitoes consist primarily of certain species of *Aedes* and *Culex*. Eggs are laid singly on the inside wall of the container, at or above the waterline, and, following a period of desiccation, hatch when inundated. Multiple broods are common. Because of the association of artificial containers with human habitation, these species may become locally abundant and annoying, and are important targets of urban mosquito control campaigns.

MOSQUITO GENERA

Most anopheline mosquitoes have spotted wings while culicines generally lack wing spots. Female anophelines have palpi about as long as the proboscis, whereas the palpi are much shorter than the proboscis in female culicines (Fig. 7.1). Anophelines rest with the head, thorax and abdomen in a straight line that is held at an angle of 40-90° to the surface on which they are sitting (Fig. 7.1). Culicines usually rest with the thorax and abdomen somewhat parallel to the surface upon which they are sitting (Fig. 7.1).

The Genus *Anopheles*

The genus *Anopheles* includes about 15 species in the United States. Members of this genus occur primarily in permanent bodies of fresh water. Eggs are always deposited singly on the water surface and are supported by lateral floats. The larval stages require from 6-7 days to several weeks for development, depending upon the species and environmental conditions, especially water temperature. Larvae feed just beneath the surface where they ingest microscopic plants and animals. Most anophelines favor water that contains aquatic plants, and larvae are often abundant among aquatic plants which provide protection from fish and other predators.

Most adult anophelines are active at night; they pass the daytime hours resting in dark, damp shelters. The night range varies from less than one mile to several miles. A blood meal is usually required before fertile eggs are produced. Species in the United States feed more frequently on the blood of domestic animals than on man. Breeding is continuous throughout the warm seasons of the year. The winter is usually passed as fertilized females. Anophelines were formerly very important as vectors of malaria in the United States.

The Genus *Aedes*

This genus contains almost one-half of all North American mosquitoes. Many of the major pest species as well as important disease vectors belong to this group. Many species of *Aedes* are vicious biters of great economic importance. Their biting habits vary but they most frequently attack during the evening hours. Some species, however, bite only during the day and others bite either by day or by night. Flight ranges are extremely variable, being rather short for domestic and woodland species and very long for some of the floodwater and salt marsh species.

Species of *Aedes* characteristically deposit their eggs either singly on the ground, on the waterline, or slightly above the waterline in tree holes or containers. Eggs withstand long periods of desiccation and typically hatch after flooding. Many of the northern species have only one brood per year; their eggs do not hatch until they have been subjected to periods of drying or cold. Other species are intermittent breeders and may have several generations per year depending on rainfall or irrigation practices. Species occurring in cold regions overwinter in the egg stage.

Larval habitats are extremely variable. In general, they are found in temporary pools formed by rains, melting snow or overflows. Some species breed in coastal salt marshes that are flooded at intervals by unusually high tides. Others occur in ponds and ditches fed by irrigation water. A few species occur in tree holes, rock pools and artificial containers.

The Genus *Culex*

About 30 species of *Culex* occur in the United States although only 12 of these are common. *Culex* species breed in quiet waters of almost all types, from that in artificial containers to large bodies of permanent water. Water in which there is considerable decaying organic material, such as animal excretement, is frequently favored. Eggs are deposited in rafts of 100 or more; they float on the water surface until hatching occurs some 2-3 days later. Breeding continues throughout the warm season with several generations a year being common in the southern states. Adult females are generally inactive during the day and bite at night.

The Genus *Culiseta*

Members of this genus are somewhat similar in appearance and habits to *Culex*. There are eight species in the United States. Two species have been found naturally infected with encephalitis virus, but their relation to the epidemiology of the diseases is not known.

The Genus *Psorophora*

Psorophora is a totally North American genus consisting of nearly 50 species, several of which were probably important vectors of Venezuelean equine encephalitis during the Texas outbreak in 1971. Breeding sites of this group are similar to those of the floodwater *Aedes*, to which they are closely related. Females deposit eggs on the ground; eggs can withstand drying and remain inactive for months, but hatching occurs quickly after flooding and the larvae develop rapidly.

The Genera *Mansonia* and *Coquillettidia*

These closely related genera include several species in the United States, only one of which is widespread and common. Eggs are deposited in rafts on protected water in marshes or lakes. After hatching, the larvae descend below the surface of the water, where they remain throughout their development, and insert their air tubes into the stems or roots or aquatic plants to obtain air. Because of this unique habit, these larvae cannot be controlled with ordinary surface larvicides.

Other Mosquito Genera

Five other genera of mosquitoes occur throughout the United States. *Uranotaenia* breed in ground pools, the grassy margins of lakes and occasionally in tree holes and pot holes. *Wyeomyia* occurs in small collections of water in the leaves of pitcher plants and bromeliads as well as in tree holes. *Deinocerites* comprises a unique genus of mosquitoes that breeds in salt water in crab holes in southern Florida and Texas. *Orthopodomyia* breeds in tree holes and in water in leaf bases of bromeliads; members of this genus also may occur in large numbers in artificial containers, along with certain species of *Culex* and *Aedes*, when organic matter is abundant. Members of the genus *Haemagogus* are mainly arboreal species that breed in tree holes and plants such as bromeliads. *Toxorhynchites* species are large, brilliantly colored, nonbloodsucking mosquitoes; their larvae are predaceous upon other mosquito larvae in tires, tree holes, bromeliads and other water-holding containers.

IMPORTANCE AND NATURE OF DAMAGE

Many species of mosquitoes are viscous biters and large numbers may attack livestock in poorly drained, irrigated and marshy areas. Such attacks result in considerable annoyance and stress to animals; blood loss and anemia may occur when extremely large populations are present. Certain species may attack domestic poultry in sufficient numbers to decrease egg production. Mosquitoes are also biological vectors of two important animal diseases: equine encephalitis and dog heartworm.

Equine Encephalitis

Mosquito-borne encephalitis is a viral disease affecting the central nervous system of horses and man. The two major types of equine encephalitis in the United States are Eastern equine encephalitis (EEE) and Western equine encephalitis (WEE) EEE occurs primarily along the Atlantic and Gulf Coasts and inland in limited areas in the Mississippi River Valley. WEE is found throughout the United States but occurs most commonly in states west of the Mississippi River and in Wisconsin and Illinois. An outbreak of Venezuelan equine encephalitis (VEE) occurred in Texas in 1971; however, its present status in the U.S. is unknown. The viruses responsible for EEE and WEE are normally transmined from bird to mosquito to bird, and less commonly from bird to mosquito to horse (Fig. 7.4). *Culiseta melanura* and other bird-feeding mosquitoes are believed to be the primary vectors in enzootic cycles of EEE while various species of *Aedes* are important in equine and human outbreaks; *Culex* and *Culiseta* are the major vectors of WEE. The occasional infection of equine hosts may result in severe illness or death. In most cases, horses are considered a "dead end" for the virus, with little or no possibility of subsequent transmission because of their inability to infect mosquitoes. The virus responsible for VEE is normally transmitted by many species of mosquitoes among small mammals, but occasionally is transmitted to horses as well. Because of the higher VEE viremias common in infected horses, these hosts may provide a source for the mosquito transmission of the virus to new hosts.

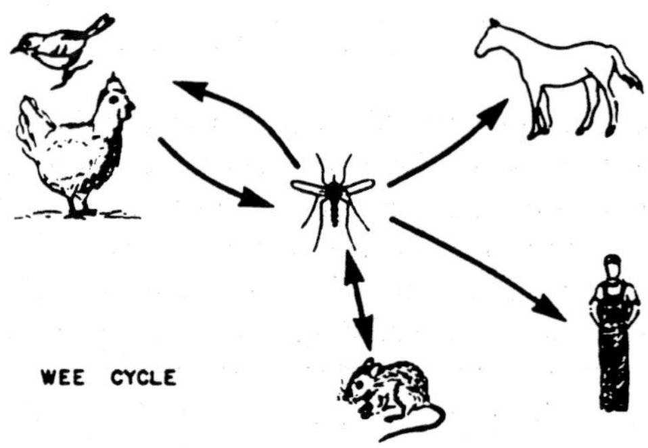

Figure 7.4. Epidemiological cycle of Western equine encephalitis (WEE). (CCEX)

Dog Heartworm

Dog heartworm, *Dirofilaria immitis*, is a common, mosquito transmitted filarial parasite of dogs and other canines along the Atlantic and Gulf Coasts from Massachusetts to Texas and inland at least to Illinois and Minnesota. Adult worms infect the right ventricle and pulmonary artery; females at least six months of age give birth to immature worms termed microfilariae. Microfilariae do not grow or develop in the canine, but circulate throughout the vascular system where they remain viable up to three years. When a mosquito (e.g.. *Anopheles*, *Aedes* or *Culex* species) bites an infected host, blood containing these microfilariae is ingested. Within the malpighian tubules of the mosquito, the microfilariae undergo development into infective larvae in 14-21 days. When the mosquito bites a dog, the infective larvae is transmitted to the dog where it developes in the intramuscular fascia and subcutaneous tissue for the next 2-3 months. They then penetrate the lumen of blood vessels and pass to the dog's heart where maturation occurs over the next 4-5 months. When fully mature. mating occurs and the female worm begins producing microfilariae and the cycle (Fig. 7.5) is repeated. Since adult worms may live as long as five years and microfilariae may persist in the dog's circulatory system for three years, a single infected dog may be a constant source of infection for as long as eight years.

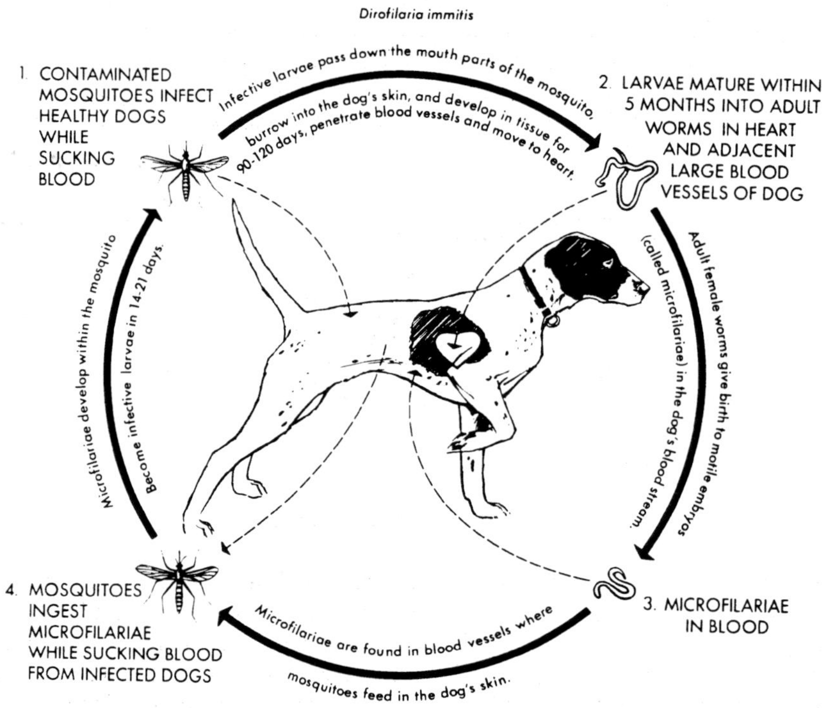

Figure 7.5. Dog heartworm life cycle.

MOSQUITO CONTROL

Many local areas in the U.S. have organized Mosquito Abatement Districts (MAD's). These agencies have technical staff and operators who survey mosquito breeding sites and conduct control operations when needed. In areas outside these districts, however, livestock and property owners must conduct their own control programs. The elimination of standing water is the most permanent method of mosquito control. Irrigated pastures should be graded and ditched to provide drainage and prevent ponding of water. Cattle should be kept off wet pastures to avoid formation of deep hoof prints suitable for mosquito breeding. Little can be done, however, for natural sources such as quiet pools in creeks and streams, or for mountain meadows flooded by melting snow run-off or from underground seepage. Water troughs can be stocked with mosquito-eating fish, *Gambusia*, which often are available from nearby control agencies. Stock ponds should be constructed with steeply sloped banks on three sides and a gently sloped fourth bank which will not be conducive to mosquito production because of animal traffic. Manure lagoons should be constructed with banks having a slope of 60 degrees. Annual weeds must be controlled around these sources as well as along the banks of drainage ditches and similar areas to reduce the number of breeding sites suitable for *Culex* mosquitoes. Consideration also must be given to potential mosquito breeding sources around residences. Openings to cesspools and septic tanks should be tightly screened and tops adequately covered with soil; open cans, old barrels and similar containers should be discarded or turned upside down to prevent collection of water. Leaking water pipes and drains should be tightened or repaired.

Control of adult mosquitoes on pastured livestock is difficult since insecticide applications provide only temporary protection of the animals. Almost daily application of repellent-type materials is required to prevent mosquitoes from attacking confined animals. Dairy cows and stabled horses present greater opportunity for daily treatment. Some dairymen install automatic spray devices which treat cows as they exit milking barns.

BLACK FLIES OR BUFFALO GNATS (FAMILY SIMULIDAE)

Members of this family, of which there are more than 1,000 known species, are small flies, 1-5 mm in length. The common name, black fly, is somewhat of a misnomer since many species are gray and others are light tan to yellow in color. Adults are typically robust with a characteristic arched or humped prothorax (Fig. 7.6), giving rise to the common name, "buffalo gnats". The family Simulidae is worldwide in distribution, extending from the tropics to the Arctic Circle. Both sexes feed on the nectar of flowers; only the female requires a blood meal for ovarian development. Although a given species may prefer a particular animal host, most will readily feed upon other host species as well. Black flies attack man and a wide variety of domestic and wild animals and birds; others feed only on cold-blooded animals. Many important pest species belong to the genus *Simulium*.

Figure 7.6. Black fly adult.

Females deposit 150-500 creamy-white eggs that darken until they are almost black just prior to hatching. The eggs usually are deposited in masses on some convenient object in or near the edge of flowing water. The water may be a slow-moving stream, but most species prefer rapidly flowing water. In rare cases, breeding may occur in marshy areas. The time required for hatching varies with the species and may be 3-5 days or as long as 30 days at low temperatures. Larvae most often are found just beneath the surface of rapidly flowing water. Most species feed on small animals, such as protozoa and crustaceans, or plants, such as algae, by straining particulate matter from water flowing by them with the aid of mouth brushes; some species feed in the silt or on submerged surfaces. Since the gill filaments of larvae extract oxygen directly from the water, they do not come to the surface to breath as do mosquito larvae. Duration of larval development, like the egg stage, varies with species and temperature; it may range from 10-14 days to 7-10 weeks. Some species overwinter as larvae; in the artic and other far northern areas, winter is usually passed in the egg stage. The last larval instar spins a reddish-brown, basket-like cocoon in which pupation occurs. These cocoons are attached to slightly submerged objects such as rocks, logs, roots and other debris. Pupae also possess respiratory filaments with which they remove dissolved oxygen from the water. The pupal period varies from 4-5 days to as long as 3-5 weeks, depending on water temperature and species. Adults emerge from the pupal case, rise to the water's suffice, unfold their wings and fly away. Mating usually occurs shortly after the initial flight. The complete life cycle, from egg to adult, varies from 6-15 weeks and the number of generations per year ranges from one to six, depending on species and climatic conditions.

There are numerous references of livestock losses along river basins in the United States, as well as other countries, resulting from black fly attacks. Large numbers of cattle, horses, mules, hogs, turkeys, chickens, sheep, dogs and cats, as well as wild animals have been killed. Death usually occurs as a consequence of an acute toxemia, caused by a vast number of bites, or as a result of anaphylactic shock. Weakness due to blood loss may also be a contributing factor to animal loss. The most common disease transmitted by buffalo gnats in the United States is a leucocytozoon infection of turkeys. This disease, caused by a protozoan, is fatal to turkeys; a similar disease also occurs in ducks. In other parts of the world, these flies serve as intermediate hosts for several nematode parasites of man and some domestic animals.

Black fly control is difficult due to the extent and number of variable larval breeding sources. Approved insecticides have been applied to rivers, streams and canals in which breeding occurs; however, pupae are not destroyed and the more effective chemicals may produce undesired environmental effects. Individual animals may be stabled during the day when populations are abundant, but this practice is not practical for large herds. Fly repellents applied to the chest, belly and ears are somewhat effective but require daily application. Muslin or cotton bags fitted over horses' ears may be used to prevent fly entry of those species attacking these areas.

BITING MIDGES (FAMILY CERATOPOGONIDAE)

Members of this family are very small, usually less than 3 mm in length. slender gnats known by such common names as punkies, no-see-ums and "sand flies", although this latter term is more correctly applied to some species of the family Psychodidae. The wings of biting midges may be clear or spotted, but they lack the dense scales of hairs that occur on the wings of mosquitoes and sand flies. Biting midges are widely distributed but large, troublesome numbers are most often found in coastal areas. Only a few of the species comprising the family are blood suckers; most of these attack a large variety of warm-blooded animals. All classes of livestock and poultry, as well as man, may serve as hosts. Most of the nonbloodsucking species feed carnivorously on small insects. Very little is known concerning the life history and habits of most of the species. Among the 50 of more genera comprising the family, *Culicoides* (Fig. 7.7) and *Leptoconops* are the most important pests of livestock; however, in certain localized areas, several other genera may be troublesome.

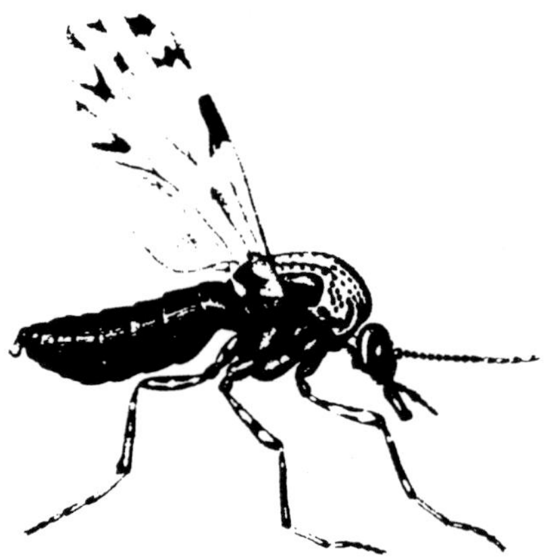

Figure 7.7. *Culicoidcs* adult

Most species oviposit on the surface of mud and water. The number of eggs deposited, oviposition site and length of time required for hatching varies with the species. Some species of *Culicoides* deposit up to 700 eggs. Temperature is an important factor in determining the incubation period which varies from 2-12 days.

Larval stages of *Culicoides* are usually found in mud, sand or debris at the edge of ponds, streams, lakes or creeks, but some species occur in decaying vegetative matter and in tree holes and under slime-covered bark. Concentrations of *Culicoides variipennis* have been found around sewage tank outlets or at the edge of stock tanks where manure is trampled into the soil. The larvae are slender and worm-like, varying from almost transparent to orange in color depending on the stage and species. The larval stage for *Culicoides variipennis* ranges from 15-23 days; some species in other genera require 8-10 months or longer to complete development. Full-grown *Culicoides* larvae usually migrate short distances from the larval habitat so that they will be above the water line and in direct contact with the air.

As with many other Diptera, only the females are blood feeders. They are extremely aggressive, crawling into feathers, hair or under clothing to get to a host. Wind and temperature appear to be the primary factors governing the activities of adults and they seldom bite in the presence of even small air currents. When the temperature drops below 55^{0} F, adult activity ceases. Animals with high body temperatures appear to be favored hosts and humans working out-of-doors are sometimes attacked in masses by these small, vicious pests. Inland breeding species produce two generations per year in most localities. Along the Gulf Coast, some species occur during every month of the year. Losses to cattle and horses caused by this group of small midges are not fully recognized; most available information is concerned with damage to man and poultry. Some species are known to prefer animals over man as hosts. In the United States, at least one species, *Culicoides variipennis*, has been incriminated in the transmission of blue tongue disease of sheep.

These gnats are difficult to control because of their extensive breeding sites. Steeply banked manure lagoons and sewage ponds do not present suitable ovipositional sites for these gnats. Chemical control of the immature stages can be achieved but is usually not economically feasible. Host treatment may offer some degree of protection, however, daily applications are necessary to prevent continuous attacks.

Eggs in mud

female - blood

blue tongue disease

steep banks

CHAPTER 8

DIPTERA: PREMISE FLIES, PASTURE FLIES, BLOW FLIES, FLESH FLIES, AND BOT FLIES

PREMISE FLIES

The premise flies, or filth fiies, consist of a number of species that are attracted to or originate from animal wastes and decaying vegetative materials associated with confinement-type animal production facilities such as feedlots, barns, stables, dairies, swine pens and poultry houses. Only a few species occur in sufficient numbers to become serious pests. The two most important species are the house fly and stable fly; however, other species such as the little house fly, latrine fly, false stable fly, syrphid flies and soldier flies may be very abundant under some circumstances.

House Fly , *Musca domestica*

The house fly (Fig. 8.1) is worldwide in distribution and clearly the most common fly species in the United States. Adults, which range in length from 6-9 mm, are dull gray in color and have four dark, longitudinal stripes on the dorsum of the thorax. Both sexes possess sponging mouthparts (Fig. 2.4) and feed on animal manure as well as secretions around the eyes and nostrils of livestock. House flies usually rest outdoors on fences and buildings but may invade homes and farm buildings in search of food and shelter. Adults appear with the first warm days of spring and remain abundant throughout the summer. Populations tend to increase during wet periods when the number and size of breeding sites increase; dry weather reduces the moisture content of spilled feed and manure, thereby reducing larval survival. The number of flies declines in the fall due to lower temperatures. House flies usually overwinter as larvae or pupae in barns and other protected sites. Breeding occurs throughout the year in southern climates, but the life cycle is greatly extended during the winter.

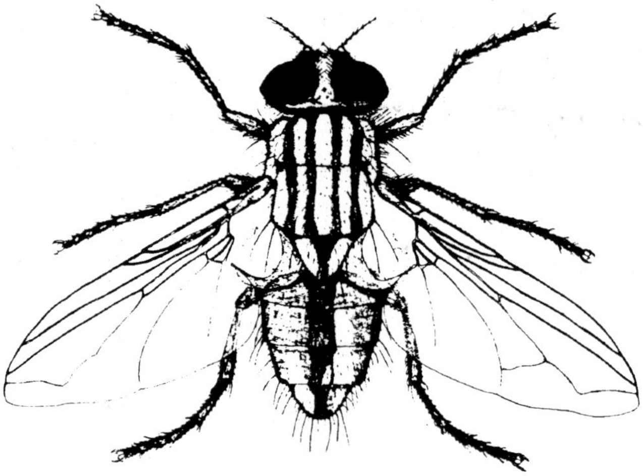

Figure 8.1. House fly adult. (PCEX)

Adult females oviposit eggs in decaying organic matter, usually spilled feed mixed with manure, or manure alone. Females deposit 100-150 eggs individually or in masses, and will lay up to 500 eggs during her life span. Eggs hatch in less than 24 hours, and larvae complete development within 5-14 days depending on temperature. Full-grown larvae seek a suitable site to pupate, usually in drier areas of the breeding material. The pupae develop into adults in 3-10 days. Females mate and begin ovipositing 2-5 days after emergence. The entire life cycle from egg to adult (Fig. 8.2) may be completed in 8-22 days.

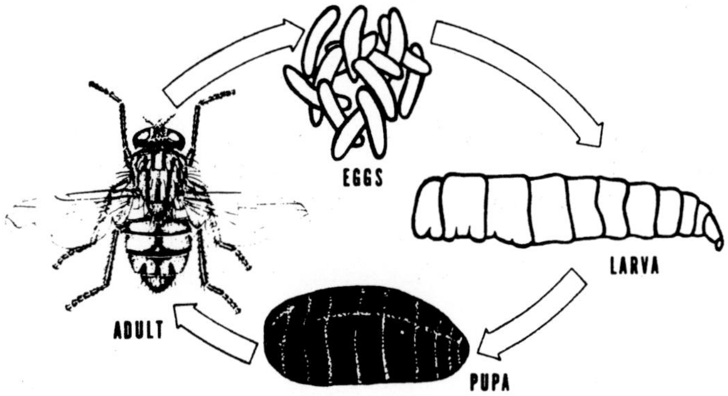

Figure 8.2. Generalized muscoid fly life cycle. (CCEX)

Livestock production losses result from annoyance and transmission of diseases such as mastitis, pinkeye, anthrax and Newcastle disease. House flies also serve as intermediate hosts of stomach worms of horses (*Habronema* spp.); cutaneous habronemiasis, or "summer sores," results when *Habronema* larvae are deposited in existing wounds by infected flies (Fig. 8.3). Large numbers of flies annoy feedlot and dairy cattle resulting in decreased weight gains and milk production. Dairymen must control flies to prevent bacterial contamination of millk. Flies may also invade nearby towns and cities and annoy residents.

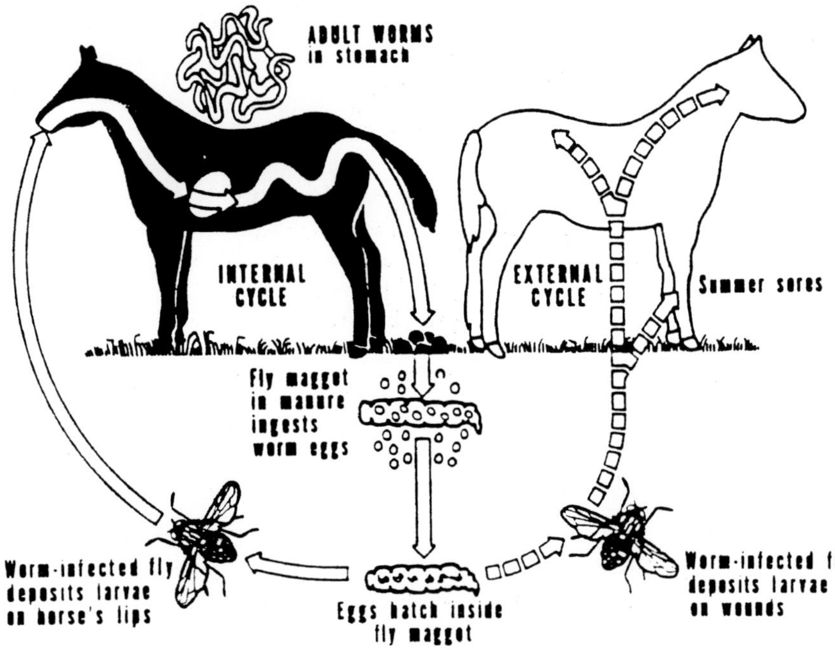

Figure 8.3. Transmission of horse stomach worm, *Habronema*, by muscoid flies. (CCEX)

Stable Fly, *Stomoxys calcitrans*

The stable fly (Fig. 8.4) is frequently mistaken for the common house fly; however, the stable fly appears brownish-gray with a greenish-yellow sheen, has a checkered abdomen, and possesses piercing-sucking mouthparts (Fig. 2.5). This species is distributed throughout the United States and most other countries of the world. Although it is called the stable fly, it is less abundant in stables than the house fly. It may enter buildings during the fall and rainy periods and will viciously attack humans, giving rise to the common misnomer of "biting house fly". The stable fly is typically an outdoor, day-biting species and is usually abundant throughout the summer and fall. Both sexes attack most warm-blooded animals, but they are commonly associated with confined animals such as cattle, horses, swine and poultry. In the northern states, overwintering occurs as larvae and pupae in moist piles of hay and manure; all stages can be found throughout the winter further south. The average life cycle requires 20-60 days, depending upon temperature and other environmental conditions. Females deposit 500-600 whitish eggs in four or five batches in moist straw, manure or other debris. Eggs hatch in 2-3 days, and larvae complete development within 2-4 weeks and pupate. Adult flies emerge 1-3 weeks later, seek hosts, mate and begin to deposit eggs. Adults usually live 3-4 weeks.

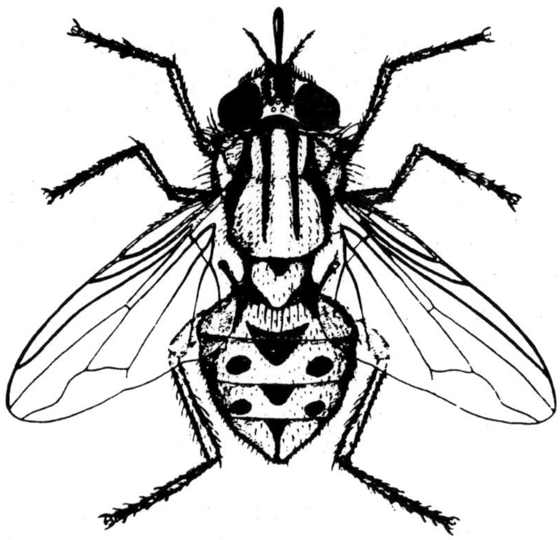

Figure 8.4. Stable fly adult. (PCEX)

Where abundant, stable flies rank as one of the more troublesome pests of confined livestock. Their annoyance interferes with normal feeding activity resulting in reduced weight gains and decreased milk production, and large numbers may cause significant blood loss. Weakened animals cease to fight flies and subsequently are attacked by ever increasing numbers. Stable flies prefer to feed on the lower legs of a host, and animals frequently stand in water for long periods to escape attack. Stable flies have also been implicated in the mechanical transmission of several livestock diseases such as equine infectious anemia and bovine anaplasmosis; in some areas, they are known to be vectors of protozoan diseases such as trypanosomiasis. Similar to house flies, stable flies may serve as intermediate hosts of the horse stomach worm; however, they are not as frequently involved in the production of summer sores as is the former species.

Little House Fly, *Fannia canniculauris*

Little house flies are small flies, seldom more than 6 mm in length. This species resembles house flies in having a dull thorax and abdomen, but differ in having only three, relatively inconspicuous dark longitudinal stripes on the thorax (Fig. 8.5). Adults are frequently observed hovering in mid air outdoors or less commonly in the middle of a room, poultry house or barn. Little house flies deposit their eggs primarily in the excrement of humans, horses, cows and poultry, or sometimes in piles of decaying grass piled up on lawns. Eggs hatch in about a day and the flattened, spiny larvae complete development within a week or two. This species is of less importance as a pest or disease vector than is the house fly; however, it may transmit Newcastle disease to poultry.

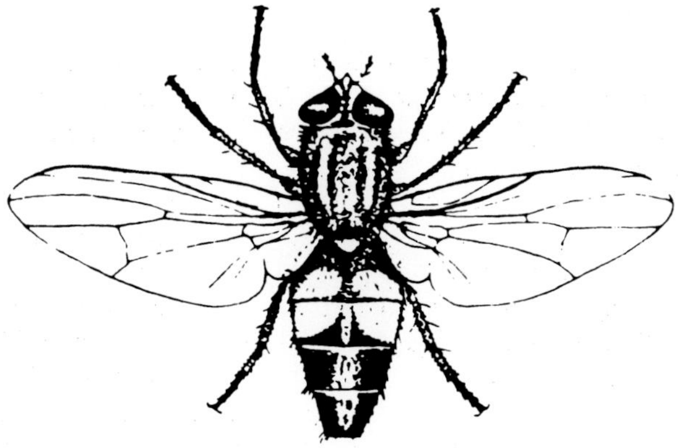

Figure 8.5. Little house fly adult. (CDC)

Control of premise flies by direct applicalion of insecticides to the host is generally not satisfactory; these species are better controlled by sanitation to prevent fly breeding. All animal wastes and decaying organic matter should be removed from the premises weekly; manure should either be properly composted or thinly spread to facilitate drying. Confinement areas should be well drained to eliminate low spots. Nevertheless, it is often necessary to supplement sanitation practices by the application of approved insecticides (e.g., space sprays and/or residual treatments) to surrounding premises. Residual sprays applied to walls, ceilings, posts and similar fly resting areas are the most common methods of treatment; space sprays or aerosols may be effective for rapid knockdown of adult flies. Care should be taken to prevent contamination of feed and drinking water with insecticides. Poison baits aid in the control of house flies, but are not effective against stable flies.

PASTURE OR RANGE FLIES

The pasture or range flies consist of a number of species which are usually associated with livestock on open range; rarely, do they become pests of livestock in confinement-type management practices. The most common members of this group are the horn fly and face fly, which are primarily parasites of cattle, the tabanids (horse and deer flies), and the sheep ked.

Horn Fly, *Haematobia irritans*

The horn fly (Figs. 8.6, 8.7) is one of the most important and economically damaging pests of range cattle in the United States. Adults are about 4 mm in length and have the same general color as the stable fly. Horn flies appear early in the spring and increase to peak populations during the late summer and early fall. Both sexes are almost exclusively parasites of cattle; however, horses, sheep, goats and dogs are sometimes attacked. Horn flies have piercing-sucking mouthparts with which they obtain blood from the host; individuals may feed as frequently as 20 times per day. Larvae develop only in fresh droppings of pastured cattle. Females dart from the host to fresh feces on which they deposit their eggs; manure apparently loses its attraction to egg-laying flies within 5-10 minutes after deposition. Eggs hatch in 1-2 days; larvae feed in the dung and become full grown in 3-5 days. Pupae form inside the puparia, either in the dung or in the soil, and adults emerge about one week later. The entire life cycle requires 10-14 days. Populations remain high until the first frost which kills most of the adult flies. Pupae overwinter in droppings or in the soil. The following spring, adult flies emerge and begin the cycle anew.

Although adults feed intermittently, they remain on the bodies of cattle day and night; they often rest on the sides and back of the animal but move to the belly during the heat of the day. Both sexes pierce the skin of the host to suck blood, causing pain and annoyance that interferes with feeding, resting and other normal activities. Heavily infested cattle may lose up to 1/2 pound of flesh per day and milk production may be reduced by 10-20%.

Figure 8.6. Horn fly adult. (PCEX)

The horn fly, due to its reluctance to leave the host, can be very effectively controlled in the adult stage by direct treatment of cattle with a number of approved insecticides. Spraying, dipping, dusting, pour-on treatments and self-application devices (dust bags, backrubbers, etc.) are common methods of treatment. Complete coverage is not essential since horn flies move about on the host sufficiently to come into contact with insecticides deposited on almost any part of the animal's body. Insecticide impregnated ear tags may provide season-long control in areas where resistance has not developed. Oral larvicides and sustained-release bolus formulations are generally effective only if all cattle throughout an extensive area are treated; such materials prevent the development of larvae in the manure but are not effective against adults.

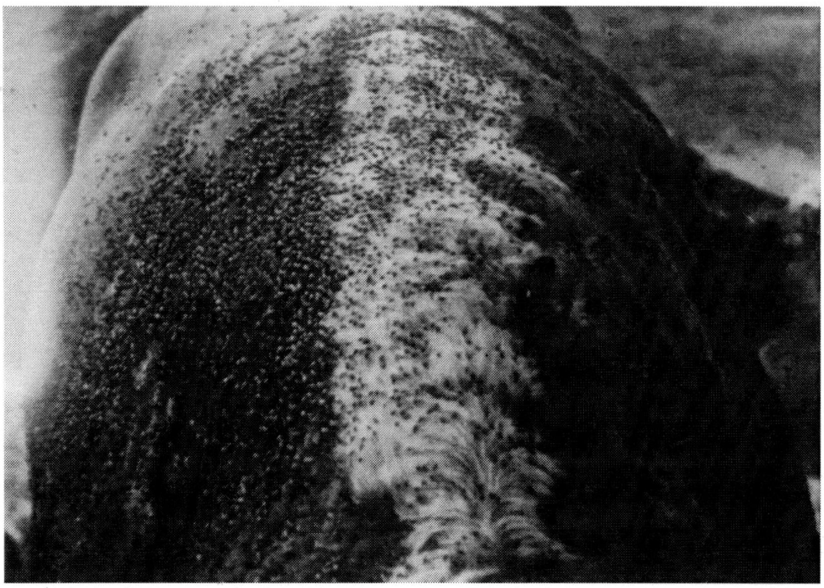

Figure 8.7. Horn fly infestation of cattle. (C. D. Schmidt)

Face Fly, *Musca autumnalis*

Face flies (Fig. 8.8) are distributed throughout the continental United States with the exception of the arid Southwest (i.e, Arizona, New Mexico and Texas). This species attacks livestock, chiefly cattle, about the head, face and neck. Although adult flies will feed on other large mammals (e.g., horses), the face fly is dependent on cattle for its survival and propagation. Similar to horn flies, female face flies deposit their eggs only in freshly excreted feces of cattle on rangeland or pasture. Eggs are not deposited in manure piles around barns and stables, or in the urine-saturated, trampled droppings usually associated with feedlots and drylots. Face flies generally are present on cattle from April or May until October, and occur in greatest numbers between July and September. Adults have sponging mouthparts. Only females are parasitic; they are found on the host for short periods to feed, taking several small meals each day over a life span of 2-3 weeks. Other times, they may be found feeding on dung or resting on nearby vegetation or on fences and buildings. Usual food sources are mucous secretions from the mouth, eyes and nostrils, and blood from injuries or wounds produced by other species of blood-sucking arthropods (e.g., horse flies, gnats, mosquitoes and ticks) (Fig. 8.9). Females are present on the host only during the day; their nocturnal habits are not well known. There is constant movement of flies on and off cattle throughout the day so that the number present on an animal at any given time may vary greatly. Males are primarily flower feeders. Adults of both sexes also feed on fresh cattle dung.

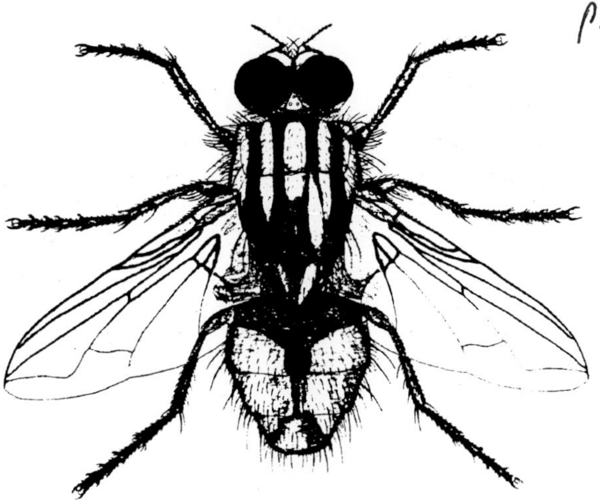

Figure 8.8. Face fly adult. (PCEX)

The life cycle from egg to adult fly may be as short as ten days during the warm summer months. Generally, eggs require from 1/2-1 day to incubate before the larvae hatch and begin to feed on manure. Larval growth and development are completed in 4 days. Pupal development may take from 3-10 days, after which time the adult flies emerge and begin to mate and feed.

The face fly is basically a pasture species, but it can become a serious domestic pest in the fall when it invades barns, homes, churches or other buildings in search of dark, protected overwintering sites. Adults mate after emerging from hibernation in the early spring; overwintered males die and the females deposit their first eggs. The second adult generation appears in 3-4 weeks and subsequent adult populations increase to maximum numbers 3-4 months later.

Heavy face fly infestations are frequently associated with an increased incidence of pink eye disease (infectious bovine keratitis) of cattle. Pink eye is an inflammation of the cornea and conjunctiva caused by several organisms but primarily by the bacterium, *Moraxella bovis*. Infected eyes lacrimate heavily; the cornea becomes clouded and the eyelids eventually close. If left untreated, affected animals injure themselves and do not perform well because of poor vision; in severe cases, blindness occurs and animals must be handled and fed separately or destroyed because of an inability to feed. The face fly is an excellent vector of this disease agent since it repeatedly feeds on eye fluids and frequently moves from one animal to another. Face flies feeding around the eyes of horses also severely irritate the tear ducts and cause excessive lacrimation but true pink eye does not occur. An eye worm, *Thelazia* sp., identified from horses in the eastern United States, is also transmitted by the face fly.

Figure 8.9. Face flies on muzzle and around eyes of a cow. (USDA)

Host treatments for face fly control are not as effective as those directed against the horn fly because face flies congregate on the face, an area that is difficult to treat, and because they characteristically spend a great deal of time off the host. Thus, residual activity of an insecticide is greatly diminished. Providing shelter for animals, although rarely practical, can be beneficial since face flies will not ordinarily enter buildings.

Tabanids (Horse Flies and Deer Flies)

Members of this family (Fig. 8.10, 8.11) are robust flies, 8 mm to 15-30 mm in length. Adults of some species are quite colorful, but most are various shades of brown mixed with yellow, gray or black. They are commonly referred to as horse flies, deer flies or green heads. Tabanids occur throughout the entire United States, but their abundance in a particular locality depends on the availability of breeding sites. Adult activity varies with the location and season. Deer flies are most abundant during the spring while horse flies predominate in the early summer and fall. Although male horse flies and deer flies do not bite; females obtain blood meals from a wide variety of warm- and cold-blooded hosts, particularly cattle, horses and deer.

Females deposit 180-1,000 spindle-shaped white, brown or black eggs in characteristically-shaped masses on vegetation growing in or overhanging swampy areas or ponds; some species oviposit on vegetation over decaying logs or concrete arches of bridges. Egg masses are usually covered with a mucilaginous substance that firmly cements the eggs together as well as to the material on which they are placed. Length of time required for hatching is influenced by temperature and species, but 4-7 days is average under summer conditions. Following hatching, larvae usually drop to the surface of the water or moist soil and quickly burrow into the mud, damp earth or decaying matter. Many species feed on organic matter while others are predators that feed on insect larvae, crustaceans, snails and earthworms; some species are cannibalistic. The larval stages are usually found in mud in marshes or along flat creek banks, roadside ditches and irrigation overflows; some species occur in rotting logs, decaying masses of leaves or even in vegetable fields. Time required for completion of larval development varies with the species and climatic conditions. Some species grow rapidly during the summer and autumn and slowly during the winter, requiring about one year for maturation; other species may complete larval development in 48 days or less and produce several generations per year.

1 generation per year

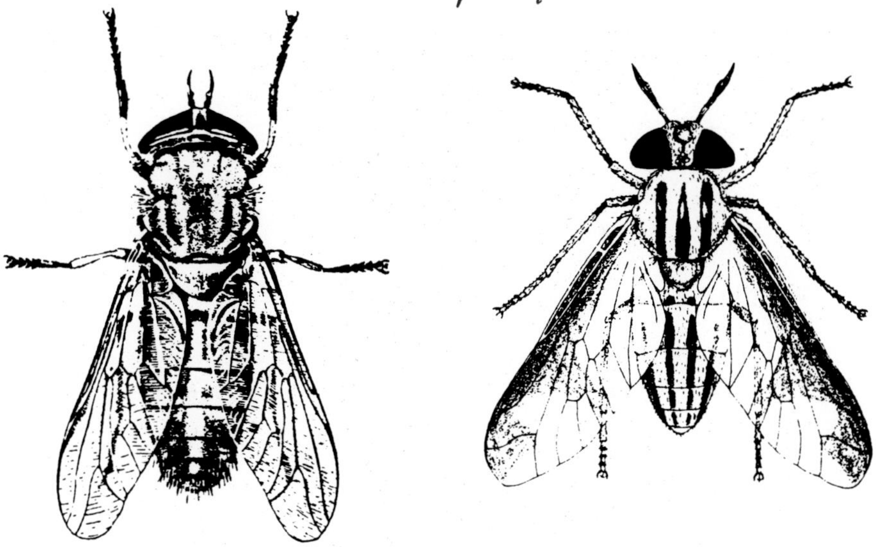

Figure 8.10. Horse fly (left) and deer fly (right) adult. (PCEX)

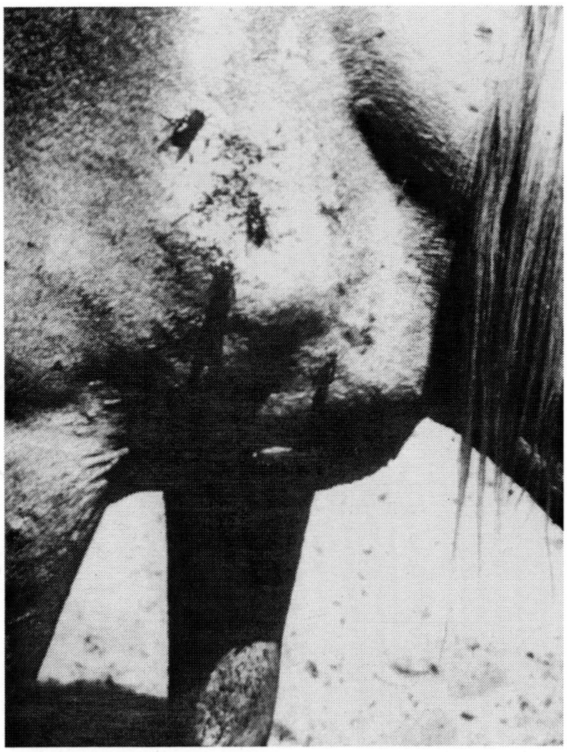

Best mechanical disease transmitter

(Tabanidae)

Anaplasmosis

EIA

Figure 8.11. Horse flies feeding on the shoulders of a horse. (USDA)

Mature larvae crawl to drier areas and pupate several inches below the surface. The pupal period varies from 2-3 weeks up to three months with temperature and species. Females of most species are avid blood-suckers and begin searching for a host soon after emergence. Males feed on pollen and nectar. Horse flies are strong, swift fliers that are most active on bright, sunshiny days when they may cover wide areas in search of a host.

Tabanids possess broad, flat, blade-like mouthparts (Fig. 2.5) that inflict rather large, deep and painful wounds. During the process of feeding, the mouthparts may be inserted and withdrawn several times which increases the annoyance, loss of blood and potential for disease transmission. The painful bite and the characteristic "buzzing sound" of the flies may cause animals to cease grazing and cluster together for protection. Such activity results in weight loss and reduced milk flow, and clustered animals may injure one another by kicking or hooking. Weight losses up to 100 pounds per animal and reductions in milk flow of 20-30% have been estimated. Another important factor is blood loss; 10-30 flies have been estimated to result in loss of almost one-third pint of blood in only six hours. Other estimates have been as high as one-third of a quart when infestations were higher. These estimates do not include the large drop of blood that is usually left at the site of each bite. The intermittent feeding behavior of these flies makes them efficient mechanical carriers of certain pathogens. A fly interrupted during feeding may pass quickly to another animal so that diseased and healthy individuals are often fed upon in succession by the same fly. Anaplasmosis, anthrax, equine infectious anemia and several other diseases are known to be disseminated by tabanids in this manner.

No satisfactory method exists for the control of horse flies and deer flies since these species are capable of flying long distances and do not stay on the host long enough to be killed by residual insecticide treatments. If possible, it is best to avoid pasturing livestock near swampy, wooded areas when tabanid populations are high. Sheltering animals may be advantageous since tabanids may enter enclosures but do not feed indoors. Drainage of low, swampy areas to eliminate breeding sites may be beneficial; however, this measure is usually not practical because of the cost involved.

Sheep Ked, *Melophagus ovinus*

The sheep ked (Fig. 8.12), sometimes incorrectly termed the "sheep tick", lives in the wool and on the skin of sheep; favored locations include the neck, breast, shoulders, belly and thighs. This parasite is a wingless fly with piercing-sucking mouthparts. Adults are reddish or gray-brown in color and measure about 6-8 mm in length. The sheep ked is widely distributed; it is more common on sheep with coarse- and medium-wool than on fine-wool breeds.

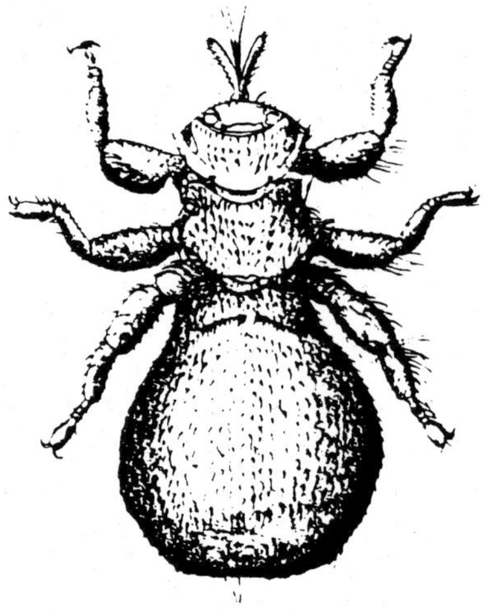

Figure 8.12. Sheep ked adult. (PCEX)

Female sheep keds deposit full-grown larvae which pupate almost immediately. The pupa, covered with a soft white membrane which hardens and turns brown in approximately 12 hours, is attached to the wool by a glue-like substance. Adult keds emerge from the pupal cases in 19-25 days. Young females may mate several hours after emergence but their ovaries are not mature until they are 6-7 days old, and the first larva is not deposited until they are 12-14 days old. Subsequent larvae are produced every 7-9 days as long as the female lives; 12-15 progeny are produced under favorable conditions.

Both sexes suck blood, resulting in considerable irritation, interference with feeding, and, consequently, poor growth and reduced vitality. Wool value is lowered due to fecal excretions and pupal case attachment. Sheep ked control is performed during the fall or winter months by spraying or dipping infested animals; however, retreatment is usually necessary within three weeks to kill young keds which have emerged from puparia after the initial treatment. Most keds are removed from sheep during spring shearing; ked-infested wool should not remain in areas to which lambs have access. Some managers pen sheep for 1-2 days until shearing wounds heal and then spray or dip these animals to prevent blow fly attack; in so doing, any remaining keds also are killed.

BLOW FLIES

Adult blow flies are often metallic in color. The most common species deposit their eggs on the bodies of dead animals and larvae feed on the decaying carcass. Some species, however, lay their eggs in open wounds of live animals; in some cases the larvae feed only on decaying or necrotic tissue of the host, but in other cases, such as the screwworm, they may attack the living tissue itself. The presence of fly larvae on a living host is termed "myiasis."

Screwworm, *Cochliomyia hominivorax*

The adult screwworm fly is approximately twice as large as the house fly; it is bluish-green in color with a metallic sheen and has three longitudinal, black stripes on the thorax. Full-grown larvae or maggots are pinkish when they are about ready to leave the wound and may be distinguished from other common blow fly maggots by two dark parallel air tubes (Fig. 8.13) that are visible through the cuticle. Screwworms have been eradicated from the United States and most of Mexico; however, the species remains well established in Central and South America. The fly continues to be a threat to U.S. livestock producers because of the danger that it might reinvade this country. Screwworms have been known to infest practically all warm-blooded animal species (i.e., mammals and birds), but infestations are most common on cattle, hogs, deer, sheep and goats.

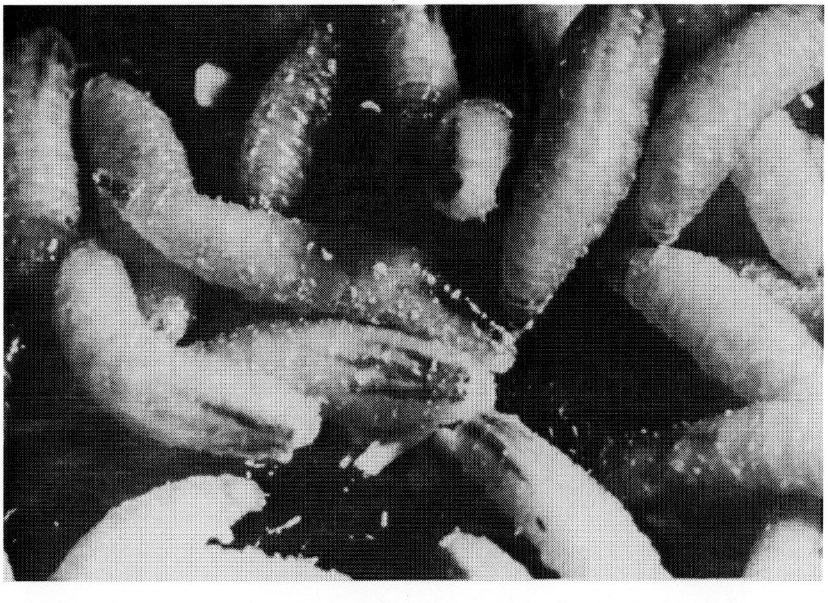

Figure 8.13. Screwworm larvae removed from a wound. (USDA)

The four stages of screwworm development (Fig. 8.14) are egg, larva, pupa and adult. Length of the developmental stages varies according to climatic conditions. Under normal summer conditions, females deposit 100-250 eggs in shingle-like masses near a wound or body opening. Blood spots remaining where ticks and biting flies have attacked animals as well as wounds caused by barbed wire cuts or scratches, dehorning, castration and peeling brands are attractive to the female fly. Eggs hatch in 10-12 hours under optimum conditions, and the small larvae enter the wound and tear out pockets of healthy flesh with their sharp mouth hooks. Maggots feed with the head embedded in the host's flesh and the posterior end exposed so that they can breath. Larvae become full grown in 5-7 days, drop from the host and pupate; adult flies emerge 7-10 days later and usually mate when they are 2-4 days old.

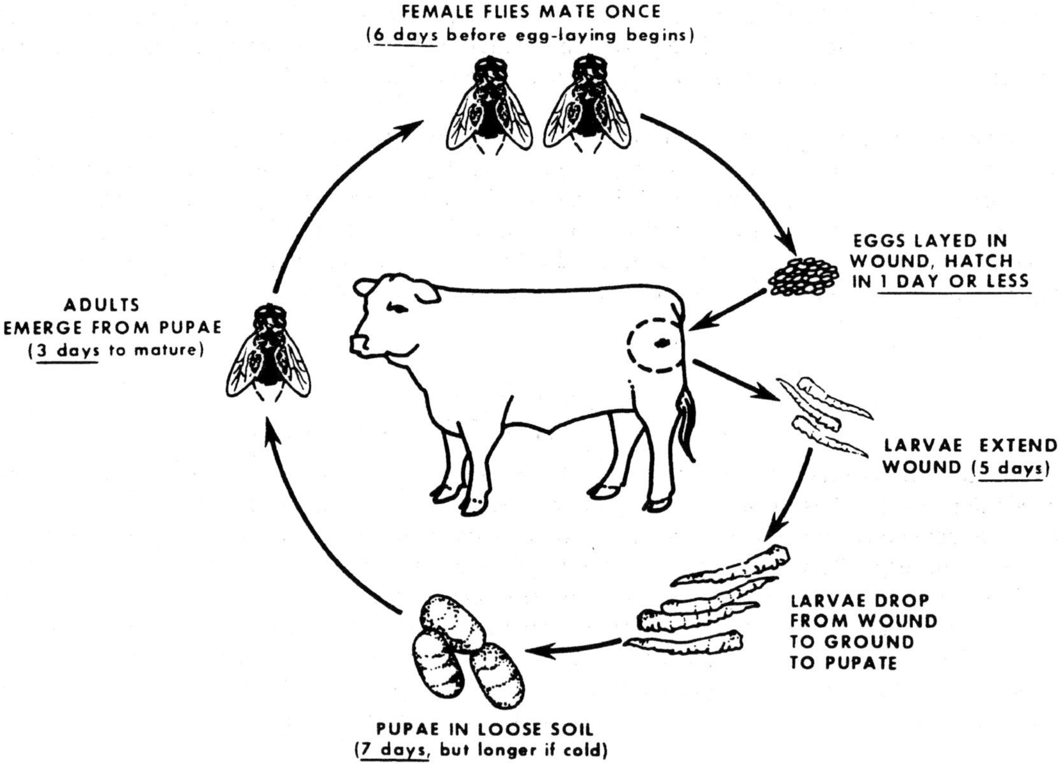

Figure 8.14. Screwworm life cycle. (CCEX)

Infested animals hide in the woods or brush, refuse to eat, and usually die if not found and treated. The odor of an infested animal attracts additional flies to deposit even more eggs so that huge wounds filled with maggots eventually result. Other species of blow flies, such as the secondary screwworm, which normally lays eggs on the carcasses of dead animals also may subsequently infest wounds initially attacked by the screwworm. Eyes, ears, tails, hooves and other body parts may be lost or destroyed. In addition to the destructive flesh-eating effect of screwworm larvae on live animals, their activity also results in pronounced weight loss, anemia and alterations of body temperature. Secondary infections are common and can cause permanent injuries such as "swollen joint" of calves. Infestations about the naval of newborn animals may cause death.

Other Blow Flies and Fleeceworms

The secondary screwworm fly, *Cochliomyia macellaria*, is a metallic green species with three black, longitudinal stripes on the dorsal surface of the thorax as in the screwworm fly, making separation of the two species difficult. However, adults of the two species can be distinguished in the laboratory. Larvae are whitish, yellowish or grayish in color, and do not have the dark, readily visible air tubes of screwworm larvae. The black blow fly, *Phormia regina*, and a greenbottle fly, *Phaenicia sericata*, are close relatives of the secondary screwworm. The black blow fly has a dark metallic, greenish-black color, without dorsal longitudinal stripes. The greenbottle fly is a brilliant metallic, bluish-green species, also lacking stripes.

While the screwworm fly breeds exclusively in the living flesh of warm-blooded animals, other blow fly species ordinarily breed either in the carcasses of dead animals or in necrotic tissue of living hosts. Forty to 250 eggs of these species are deposited in rounded, irregularly flattened masses loosely cemented to the skin near an attractive wound on the host. Egg masses may be composite deposits of several species so that a suitable location may contain thousands of secondary screwworm and other blow fly eggs. Eggs normally hatch in a day or less, and the larvae become full grown in 6-10 days. Full grown larvae leave the carcass or wound and burrow into the soil and pupate. The pupal stage lasts 3-5 days, depending upon weather conditions. The total developmental period varies from 9-39 days. All three species commonly infest sheep with wool that is soiled with urine and feces; maggots feed on the soiled wool causing it to loosen and putrefy, giving rise to the common name of "fleeceworms" or "wool maggots". Unless treated promptly, infested sheep may be quickly killed.

Many of the more common causes of screwworm and blow fly attack can be prevented by timely livestock practices that minimize susceptibility. Dehorning, castration and branding should be performed so that wounds will not be available to flies during their season of abundance. Infested wounds must be cleaned and treated with an approved insecticide; some of the best preparations contain an antiseptic or antibiotic to facilitate healing. Extension personnel and/or a local veterinarian should be notified if screwworm infestation is suspected. Sheep infested with wool maggots should be sprayed or dipped with an approved insecticidal solution.

FLESH FLIES

Adult flesh flies are medium in size and usually grayish in color, often with a "checkerboard" pattern on the abdomen. Similar to blow flies, many species of flesh flies breed in animal carrion and other decaying organic matter; however, some members of the group are parasitic on other arthropods. Females give birth to small larvae rather than depositing eggs. The red-tailed flesh fly, *Sarcophaga haemorrhoidalis*, is a widely distributed species that is typical of this group. Larvae normally feed as scavengers but may occasionally be found in the necrotic tissue of wounds of living animals. The life cycle requires not more than 14-18 days. Host treatment is similar to that for blow fly infestation. *Wohlfahrtia* species occur throughout the cooler regions of North America. Skin lesions are the usual site of larval entry, but the mucous membranes of the head openings and genital orifices may also be used. Domestic animals attacked include dogs, horses, cattle, goats, swine, sheep and poultry.

BOT FLIES

Adult bot or warble flies are large, stout-bodied, hairy flies that superficially resemble bees in appearance. Larvae are parasitic in various species of animals. Unlike blow flies, however, bot flies attack healthy animals; a previous wound is not necessary to attract the egg laying female. Adults have vestigial mouthparts and are incapable of feeding. Most species of bot flies have only one generation per year.

Cattle Grubs, *Hypoderma species*

The common cattle grub, *Hypoderma lineatum* (Fig 8.15), is the larval stage of a bot fly often referred to as the heel fly; it occurs in all states except Alaska but is rare in the extreme southern United States. A related species, the northern cattle grub, *Hypoderma bovis*, is abundant in southern Canada and the northern United States. Common cattle grub populations vary tremendously within their respective ranges due to diverse environmental conditions; populations within a specific area also fluctuate considerably from year to year. Cattle are the only important hosts; preferred hosts appear to be yearlings, calves and older animals, in that order. Horses are sometimes infested, but there is no record that grubs complete development under these circumstances.

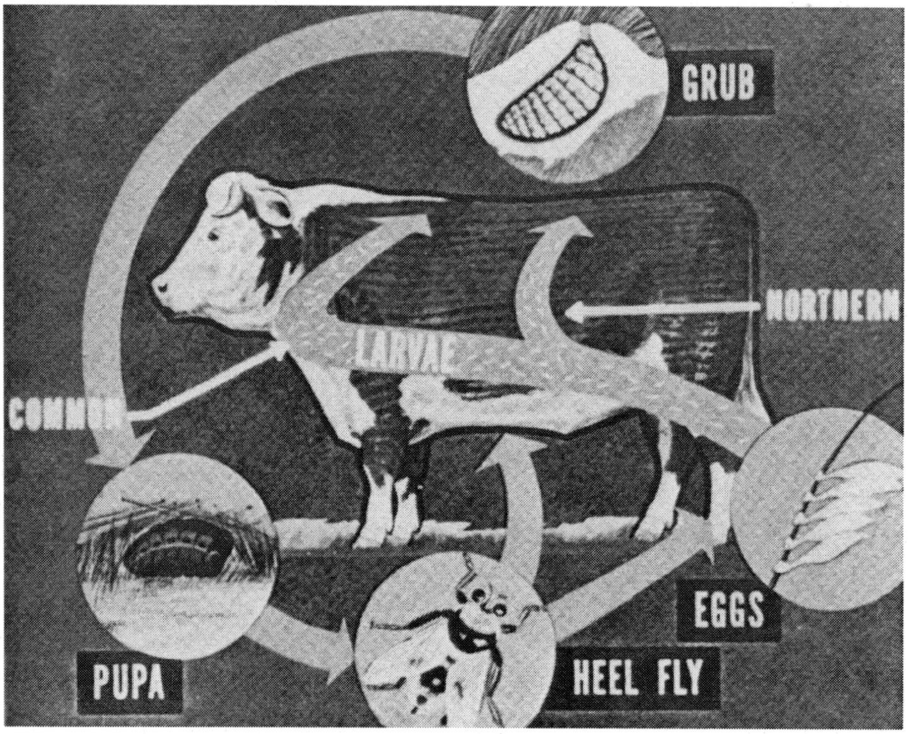

Figure 8.15. Life cycle of cattle grubs via the esophagus by the common cattle grub and the spinal canal by the northem cattle grub. (CCEX)

The non-feeding, sexually mature adult fly, about the color of a honey bee and nearly twice as large, emerges from the puparium during warm, sunny days of late winter and early spring. Mating occurs within an hour after emergence, and females deposit fertile eggs 20 minutes later. During favorable weather, flies usually exhaust themselves and die within 2-3 days; however, some survive as long as 3 1/2 weeks if cool weather restricts their activity. Common cattle grub adults are less aggressive in their ovipositional activity than the northem species, and "gadding" (Fig. 8.16), the chatazteristic reaction of cattle fleeing from ovipositing flies, is less dramatic. Common cattle grub adults most often approach a reposing animal and attach their eggs to hairs on the legs or belly near the ground without noticeably disturbing the animal; it is not uncommon, however, to see animals fleeing wildly or standing in water up to their bellies or in dense shade to escape heel fly attack. Gadding appears to be an instinctive reaction on the part of the host since the adult fly, which neither bites nor stings, causes no pain during this egglaying activity.

Eggs of the common cattle grub are attached in rows of 2-3 to as many as 25-30 along a single hair (Fig. 8.15); the northern cattle grub characteristically deposits only a single egg per hair. Females usually produce 400-800 eggs with an average of 500 eggs per fly. The eggs hatch in 2 1/2-6 days, and the larvae crawl down the hair shaft and penetrate the skin, usually at or near the base of the hair follicle, and begin an extensive migration through the connective tissues of the host to eventually encyst in the back. Mortality of first-instar larvae probably exceeds 50% and may approach 100% in highly resistant animals. First stage larvae spend approximately eight months migrating through the connective tissues of the host; movement is assisted by enzymatic secretions. Common cattle grub larvae feed in the submucosa of the esophagus (gullet) during a portion of this time, but they also have been found in the diaphragm, pericardium, spleen and rumen (Fig 8.15). Northern cattle grubs, on the other hand, typically encyst in the epidural fat along the spinal canal (Fig. 8.15). Upon reaching the back, the first stage larvae cut breathing holes in the skin, reverse their position and form a small pouch, termed a warble, in which they undergo the remainder of their development (Fig. 8.15). Growth during the second larval stage is rapid as the grub feeds on pus, necrotic cells and secretions from the wall of the warble. After several days of feeding, larvae molt to the third stage which is initially tan but appears nearly black when full grown. After developing 1-2 months in the warble, the full-grown grub squeezes through the breathing hole in the skin, drops to the ground and seeks shelter under leaves or other debris. Within 2-3 days, the larval skin hardens into a puparium, inside which the adult fly develops. The pupal stage is usually completed in 20-60 days depending on temperature. Only one generation per year is completed.

Figure 8.16. "Gadding" of cattle to escape heel fly attack. (USDA)

Warbles in the back reduce carcass value due to tissue inflammation, and breathing holes in the skin decrease the hide's leather value. Gadding may cause decreased weight gains and lowered millk production since animals are unable to graze normally.

Cattle grubs are readily controlled by treatment of the host with an approved systemic pour-on or spot-on insecticide; however, timing of application is important. For most effective results, cattle should be treated as soon as possible after adult activity ceases. Adverse host-parasite reactions such as bloat, salivation, staggering and paralysis may occur if animals are treated when grubs are in the gullet or spinal cord; cattle should be treated either before or after these stages of development.

Horse Bots, *Gasterophilus species*

Three species of horse bots occur in the United States. The common bot fly, *Gasterophilus intestinalis*, and the throat bot fly, *Gasterophilus nasalis*, are distributed throughout the United States; the third species known as the nose bot fly, *Gasterophilus haemorrhoidalis*, is more common in northern and midwestern states. Horse bot flies have four stages of development (Fig. 8.17): adult, egg, larva (bot), and pupa. Adult flies do not feed; they live from a few days to three weeks.

Adult flies appear as early as March and are present throughout the summer into late fall. Eggs, deposited singly and attached to hairs on various parts of the host's body, usually can be found soon after flies are first observed in the spring. The common bot fly deposits eggs primarily on the forelegs of the horses, the throat bot fly attaches eggs to hairs under the jaws, and the nose bot fly deposits eggs at the base of hairs around the lips. Females deposit from 160-500 eggs each. Eggs normally hatch in a few days to two weeks; however, hatching may be delayed 2-3 months under adverse conditions. Larvae of each species vary in habit, but generally the newly-hatched larvae crawl into the animals mouth or are licked off the body. They burrow into the tongue, inner lips or the gums between the teeth. Larvae remain in the mouth undergoing development for 3-6 weeks, after which time they pass into the stomach where they attach and complete development for several months. Full-grown larvae of the common bot fly and the throat bot fly pass out of the animal with the feces; nose bot larvae reattach in the rectal area for 2-3 days. Larval development of the three species usually requires 10-11 months. After leaving the host. the larva pupates for two weeks to two months. The adult fly then emerges, mates and deposits eggs.

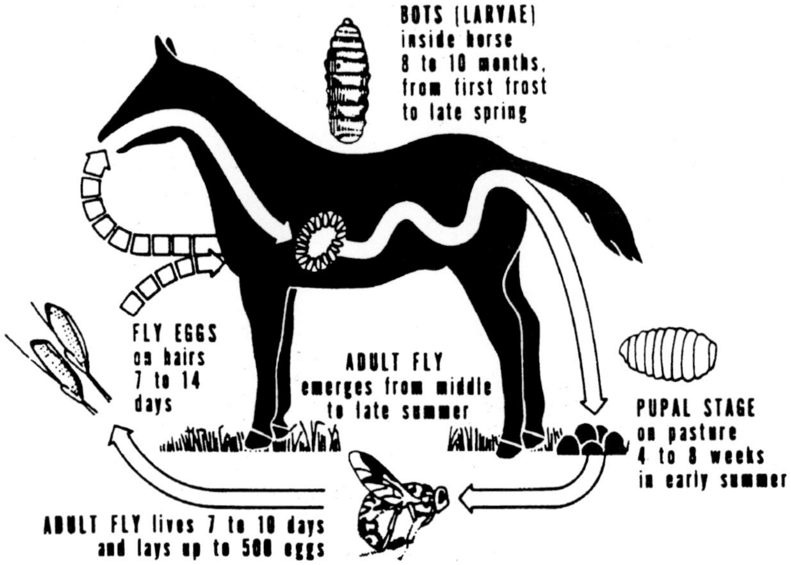

Figure 8.17. Common horse bot life cycle. (CCEX)

The common bot fly is the least annoying of the three species, probably because it does not fly about the animal's head as do throat and nose bot flies. On warm, sunny days when bot flies are most numerous, horses may be observed constantly bobbing their heads as they walk; occasionally they may break into a run, attempting to flee the flies. Animals may gather in the shade on high ground when the wind blows or enter barns of other shelters. While horses are fighting flies, they are unable to graze and may subsequently lose weight and suffer from lack of proper nourishment. Newly hatched bots severely irritate the horse's tongue, gums or lips as they burrow into these tissues. Horses frequently rub their lips on the ground, fences, stones or other objects in attempts to relieve this irritation. Bots attached to the lining of the stomach or small intestine result in inflammation which interferes with digestion; attachment to the lower part of the stomach may interfere with food passage. Infested animals often suffer from colic and other gastric disturbances. The degree of damage caused by feeding larvae is proportionate to the number present. Several hundred larvae may occur in an animal.

External treatment of the host may be initiated when eggs are first observed by applying a solution of an approved insecticide; the egg-infested area should be briskly rubbed or brushed to stimulate hatching so the young larvae will be killed by the insecticide. It is usually sufficient to treat once a week in this manner. Internal treatment may be readily performed by the addition of systemic insecticides to feeds. Other methods of application involve the use of a stomach tube or balling gun. Medical attention for post-treatment symptoms such as colic, diarrhea or ataxia may be required. Treatment should be performed 2-4 weeks following the first frost that ceases adult fly activity.

Sheep Bot Fly, *Oestrus ovis*

The sheep bot fly is the adult of the "head grub" or "head maggot" of sheep. The species is world-wide in distribution and is found throughout most of North America Larvae infest the nasal cavities and frontal sinuses of sheep and, occasionally, goats. The adult fly is most active during the heat of the day from late spring into early fall. The female fly deposits a tiny larva, or grub, on or near the nostril of sheep. The grub enters the nostril, gradually moves up the nasal passage, and usually enters a sinus cavity where it develops until full grown. During the summer, development may be completed within a month; grubs that enter the nostrils in the fall apparently remain in the nasal passages throughout the winter, move to the sinuses the following spring or summer, and then complete development. When full grown, the grubs leave through the nasal passages, fall to the ground and pupate. After three weeks to two months, the adult flies emerge from the pupal cases and seek new hosts. Under favorable conditions, the entire life cycle (Fig. 8.18) may be completed in 2-3 months.

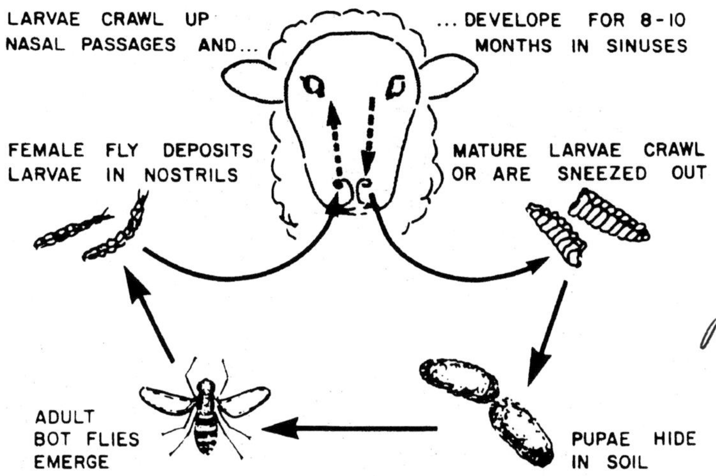

Figure 8.18. Sheep bot life cycle. (CCEX)

As the grubs crawl about in the nostrils, they induce considerable inflammation of the mucous membranes and excessive nasal secretions, a condition commonly referred to as "snotty nose." Secondary bacterial infections may cause these secretions to thicken to the extent that interference with breathing and impairment of the animal's general health follow. The sinuses may fill with pus and toxic substances may be absorbed into the host's system. Grubs sometimes die in the sinuses because they have become too large to exit; dead larvae harden in the sinus and cause abscesses to form. The behavior of sheep during fly activity is characteristic. Animals under attack stand with their noses pressed into each others fleece or into the ground; they also shake their heads and stamp their feet in efforts to prevent deposition of larvae on their nostrils. Presence of the maggots may cause giddiness or "blind staggers." Infested sheep have difficulty breathing and sneeze frequently; the eyes may become inflamed and the head may be carried low or moved about in a peculiar way in an effort to get rid of the obstruction. Animals lose condition due to lack of appetite or distraction from normal grazing. Convulsions and death may occur in severe cases. Larvae sometimes migrate into the horn cores but, contrary to popular belief, it is impossible for them to enter the brain cavity. As many as 80 larvae have been collected from one head, but the usual number is 5-12. There is no approved insecticidal treatment for bot infested sheep.

CHAPTER 9

TICKS

CLASSIFICATION

Ticks, which belong to the Class Arachnida, differ from insects (Class Insecta) in that (1) the head, thorax and abdomen are all fused into a single body region; (2) antennae are absent and (3) four pairs of legs are present in the nymphal and adult stages (larvae have only three pairs of legs). The suborder of ticks (Metastigmata) differs from suborders of mites in that (1) the tarsus of the foreleg has a conspicuous sensory pore, termed Haller's organ, and (2) most species have a prominent, toothed hypostome which is absent in mites. There are two main groups (Families) of ticks: the hard ticks (Ixodidae) and the soft ticks (Argasidae).

Hard ticks are readily distinguished from soft ticks by the presence of a dorsal shield, or scutum, immediately behind the capitulum. The dorsal shield is small in the female (Fig. 9.1); but, it covers the entire dorsal surface of the male. Soft ticks do not have a dorsal shield (Fig. 9.1). The capitulum, which bears some resemblance to a true head, extends forward from the anterior end of the body in hard ticks (Fig. 9.1), while in soft ticks it is found on the ventral side and is not usually visible from above (Fig. 9.1). When eyes are present in hard ticks, they are located on or near the lateral angles of the scutum (Fig. 9.1); in soft ticks, they are on the sides of the body above the bases of the second and third pair of legs. Hard ticks typically take one blood meal in each of the three developmental stages (larva, nymph and adult); both sexes are blood feeders, but only the female becomes greatly distended during engorgement. Most species feed on a different host during each stage, but a few species feed on only one or two hosts. Soft ticks are typically multiple feeders during the adult stage. In the United States hard ticks are much more abundant than soft ticks, cause greater annoyance, and are far more important in the transmission of disease to man and animals.

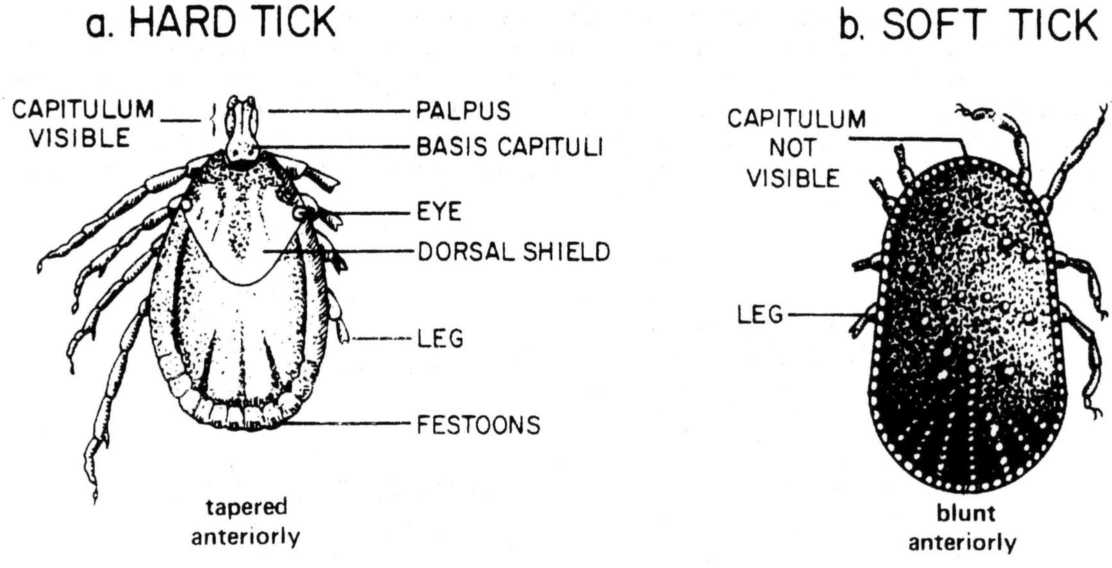

Figure 9.1. Generalized dorsal views of female hard and soft tick. (CDC)

MORPHOLOGY

The capitulum of ticks (Fig. 9.2) consists of a basal portion, the basis capituli (Fig. 9.2), to which the hypostome, chelicerae and palps are attached. The hypostome (Fig. 9.2) bears many rows of recurved barbs which anchor the tick to its host. The chelicerae (Fig. 9.2) serve as cutting agents that permit insertion of the hypostome. Palps (Fig. 9.2) have four segments, with the fourth segment very small in hard ticks but approximately equal in size in soft ticks. After the chelicerae pierce the skin of the host, the barbed hypostome is thrust deeply into the wound; the palps do not penetrate the skin of the host. The salivary glands of ticks produces a secretion which prevents blood from coagulating. Blood is pumped in by a muscular pharynx and forced backward into the esophagus and midgut. Diverticula attached to the midgut become greatly distended to store a large quantity of blood.

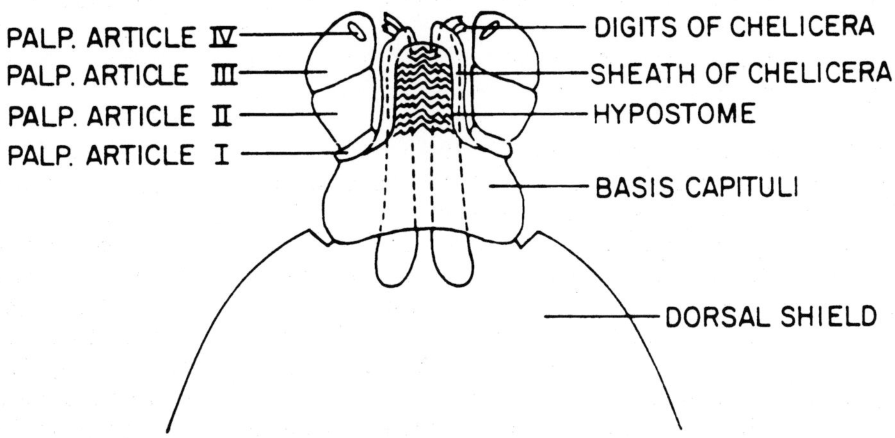

Figure 9.2. Capitulum of a hard tick. (CDC)

The outer body covering (cuticle) is leather-like and capable of great distention during feeding. Unengorged female ticks are somewhat flattened and tapered anteriorly, but become enlarged as much as twenty fold following feeding; male hard ticks take only small blood meals and increase in size only slightly. The genital pore of the male and female is situated on the ventral side, posterior to the mouthparts, between the bases of the second pair of legs (Fig. 9.3). Genital grooves extend backward from the genital pore to the posterior margin of the body in most species. The anus (Fig. 9.3) is situated between the genital grooves far behind the fourth pair of legs, near the posterior end of the body. Spiracles are borne on plates (Fig. 9.3) just posterior to the bases of the fourth pair of legs in hard ticks and on the sides of the body between the third and fourth pair of legs in soft ticks. The spiracular plates may bear finely- or coarsely-shaped punctures termed goblets. Spiracles are absent in larval ticks; respiration is transcuticular.

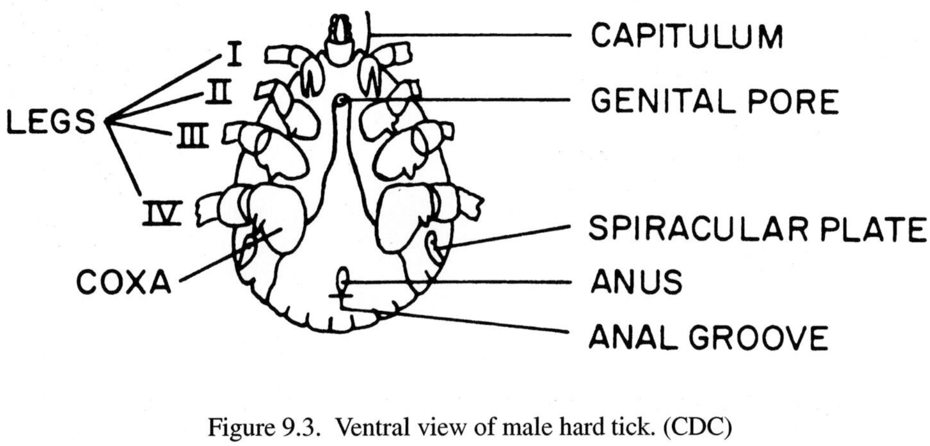

Figure 9.3. Ventral view of male hard tick. (CDC)

GENERALIZED LIFE CYCLE

Ticks reproduce sexually. The female has a duplex ovary, uterus and a vagina which leads to the external orifice or genital opening. The male has a similar reproductive system, with testes, vas deferens and external opening, to produce capsules, or spermatophores, which enclose the numerous sperm cells.

Ticks undergo four stages in their development (Fig. 9.4): egg, 6-legged larva, 8-legged nymph and adult. Larvae, or "seed ticks", possess only six legs and are not distinguishable as to sex. Host attachment is precarious and larvae use stored food to survive while seeking a host. Despite tolerance to starvation, a very high percentage fail to survive. Larvae often climb up on vegetation, waiting for a host to pass within reach; others actively seek hosts. Following attachment and feeding, most engorged larvae usually drop to the soil and molt to the 8-legged nymphal stage; however, larvae of one- and two-host ticks remain on the host to molt. Depending upon the type of development, the nymph may also detach from the host before molting. The adult must also undergo a critical waiting period for a suitable host. Hard ticks have only one nymphal instar, the nymph becoming an adult after molting; soft ticks may have several nymphal stages. Sex usually is distinguishable for the first time as the adult female hard tick differs from the male in having a small scutum. The sex of soft ticks may be determined by the shape of the genital opening located between the second pair of legs. In male soft ticks, the genital opening is almost circular while it is oval, definitely broader than long, in females. Both male and female hard ticks are blood feeders, engorging for several days before mating. After mating with one or more females, the male hard tick drops to the ground and dies. The female also drops to the ground after feeding, oviposits and dies. Female soft ticks, on the other hand, may deposit several small batches of eggs but a blood meal is usually required prior to each episode of oviposition.

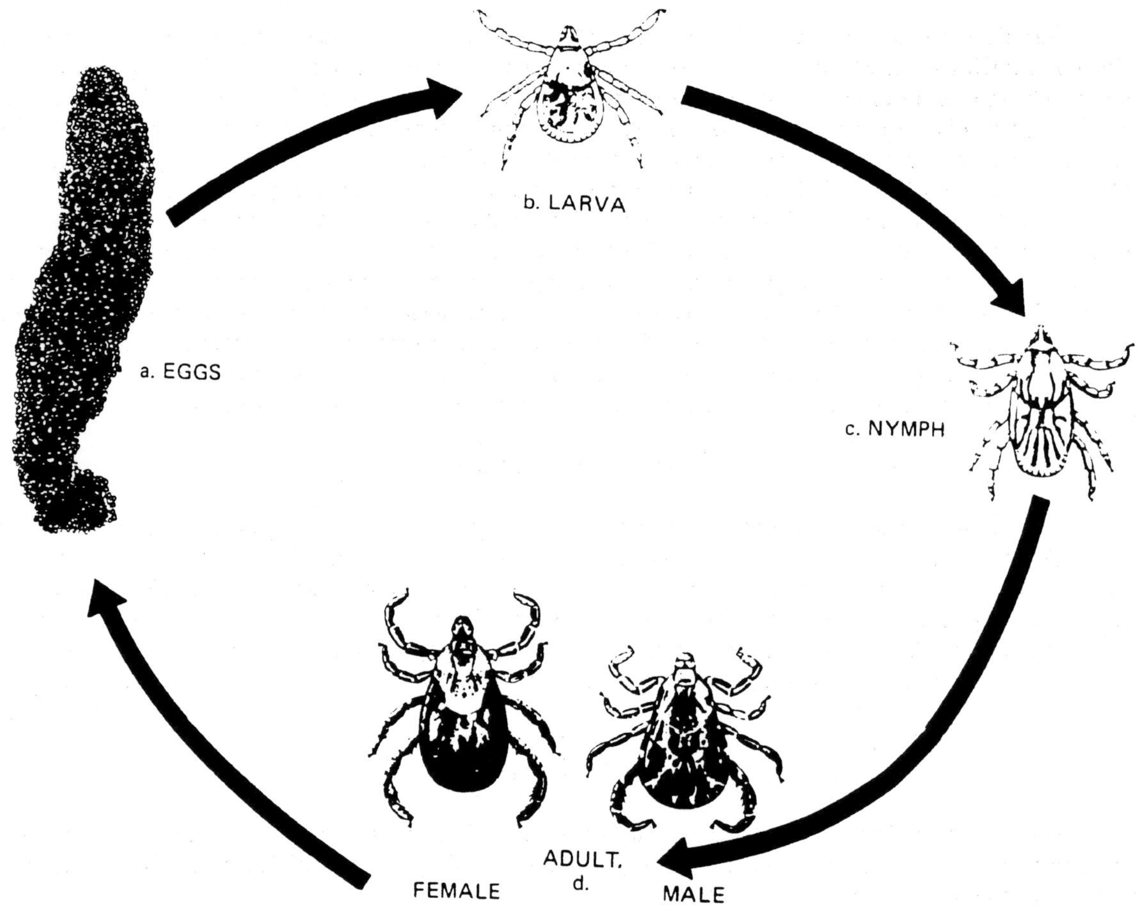

Figure 9.4. Hard tick life cycle. (CDC)

There are several types of lifecycles, depending on the number of hosts involved, by which ticks are characterized. In one-host ticks such as *Boophilus annulatus*, all the parasitic stages, beginning with the unfed larvae, are passed on a single host. They molt twice on the host and do not leave until the fully engorged, fertilized female drops to the ground to oviposit. Larvae of two-host ticks attach to one host, feed, molt to the nymphal stage and engorge, after which they detach and molt to the adult on the ground; adults later reattach to a new host to feed. A tick which requires three hosts to complete its life cycle is termed a three-host tick, as exemplified by the Rocky Mountain wood tick, *Dermacentor andersoni*. This species requires a small mammal for the larval stage, another small mammal for the nymph and another usually larger mammal for the adult; all stages detach from the host and molt on the ground. In the multi-host tick life-cycle, as in some species of soft ticks, a large number of hosts may be required; it is common to have five molts, each completed after engorging on and detaching from the host. Further, the adult stage may feed intermittently on several successive hosts. Each time a tick leaves its host, it risks its survival on the chance that it will find another host. Species which utilize more than one host are able to exist because of their high reproductive capacity and their ability to survive for long periods of time without food. Many ticks require a year or two to complete their life cycle, but this time interval may be extended under unfavorable conditions. Under optimal laboratory conditions, the usual two-year life cycle of *Dermacentor andersoni* and *Amblyomma americanum* can be reduced to 2.5 and 3.5 months, respectively. Some ticks are long-lived. Specimens of *Ornithodoros turicata* have survived in the laboratory for more than 12 years. *Ornithodoros tholozani*, an Old World species, is known to have a life span up to 20-25 years. The ability of many ticks to survive long periods away from a host is an important factor complicating the control of ticks and tick-borne diseases.

When ticks are not on a host, they are generally found on the ground or near the ground in vegetation awaiting contact with another host and the opportunity to resume their parasitic existence. The length of time a tick remains attached to the host for feeding depends on the species and the stage of its development. Hard ticks usually remain attached to a host for several days or weeks. Many of the soft ticks, however, attach to a host only briefly, perhaps an hour of less, and detach as soon as they have fed. At other times, they remain secluded in the immediate vicinity of the host's place of habitation. Climatic factors, particularly temperature, are important determinants of tick development and activity. Species vary in their ability to withstand temperature extremes; some overwintering as adults, nymphs or larvae, while others are capable of surviving high temperatures and arid conditions. Some hard ticks, such as those in the genus *Boophilus*, may produce up to five generations per year under favorable climatic conditions. The full cycle of *Dermacentor andersoni*, on the other hand, may require up to three years at higher altitudes or at the northern limits of its range. The winter tick. *Dermacenteror albipictus* is inactive during the summer months, becoming active and seeking hosts with the advent of cold weather. However, the higher temperatures of spring and summer do accelerate the development and activity of most species.

HOST SPECIFICITY

Ticks parasitize a wide range of host animals. Only a few species show a marked preference for feeding upon a single host species. Ticks feed on mammals, birds, reptiles and some amphibians. During feeding, the tick inserts the barbed hypostome into an incision made by its chelicerae. In the hard ticks, only females become greatly distended when engorged; both sexes of soft ticks become distended. *Ornithodoros hermsi* engorges in 15-20 minutes while the American dog tick and most other hard ticks feed for 10-12 days. Ticks consume large quantities of blood; severe infestations result in anemia, weight loss and even death. Other losses include those caused by "tick worry" and hide damage.

ECONOMIC IMPORTANCE

Ticks serve as biological vectors of a number of pathogenic organisms through transstadial or transovarian disease transmission. Infected ticks may require several hours of feeding before sufficient numbers of pathogens are passed to a susceptible host to result in infection.

The tick-borne diseases of domestic animals are usually recognized as being caused by an organism classified in one of the following groups: protozoal, rickettsial, bacterial and viral. Bovine babesiosis, commonly called cattle tick fever, is produced by *Babesia bigemina*, a protozoan parasite of red blood cells; principal vectors include *Boophilus annulatus* and *Boophilus microplus*. As the parasites multiply, the red blood corpuscles are destroyed resulting in hemoglobinuria and anemia. Similar parasites transmitted by *Dermacentor nitens* and *Rhipicephalus sanguineus* infect horses and dogs, respectively. Several protozoal parasites of the genus *Theilera* cause theilerioses, or East Coast fever, of cattle in eastern and central Africa; benign strains of *Theileria* spp. have been reported from cattle and deer in the United States. The rickettsial parasite, *Anaplasma marginale*, which causes bovine anaplasmosis, also multiplies in the red blood cells of cattle; death of the host may result from severe anemia. Although ticks are not the only carriers of this disease, *Dermacentor andersoni* and *Boophilus microplus* are suspected vectors. Canine ehrlichiosis, or malignant canine rickettsiosis, is a septicemic disease of dogs caused by the rickettsia, *Ehrlichia canis*; *Rhipicephalus sanguineucs* is the primary vector. Cattle and sheep each have a benign form of ehrlichiosis caused respectively by *Ehrlichia bovis* and *Ehrlichia ovis*. Bovine ehrlichiosis is common in South Africa where the vectors are *Hyalomma* spp.; the disease has also been reported from the United States. Ovine ehrlichiosis has been reported from Africa where the principal vector is *Rhipicephalus evertsi*. Heartwater is a rickettsial-like disease, caused by *Cowdria ruminantium*, that affects sheep, goats and cattle; the disease is common in parts of Africa where it is transmitted by several species of *Amblyomma*. Tularemia, caused by *Francisella tularensis*, is primarily a disease of rabbits and rodents which may be transmitted by several species of *Dermacentor* and by *Haemaphysalis leporispalustris*. Although ticks are not known to be involved in the transmission of *Brucella abortus* or *Brucella melitensis*, the causative agents of brucellosis, in the United States, there are reports which incriminate their involvement in other areas of the world, including Mexico. Spirochetosis is a mild febrile disease reported from cattle, sheep, goats and horses in South Africa; the causal agent, *Borrelia theileri*, is transovarially transmitted by *Boophilus decoloratus* and *Rhipicephallus evertsi*. *Borrelia coriaceae*, the causative agent of epizootic bovine abortion ("foothill abortion"), is transmitted by the pajaroello tick, *Ornithodorus coriaceus*, in California and Oregon. *Ixodes dammini* is the principal vector of Lyme disease, an arthritic condition caused by the spirochete, *Borrelia burgdorferi*. Nairobi sheep disease is an infectious viral disease of sheep and goats that occurs in eastern and central Africa; the primary vector is *Rhipicephalus appendiculatus*.

Man, domestic animals and birds are subject to a phenomenon termed tick paralysis in North America, Europe, Africa and Australia. A toxin secretion, produced by the salivary glands, results in an ascending paralysis that usually leads to incoordination and collapse within 4-7 days of attachment and feeding by the tick. In many instances, symptoms may subside quickly when ticks are removed if paralysis has not progressed too far. If ticks are not removed, affected hosts may die. Tick paralysis is primarily produced by engorging females; the tick most frequently involved in North America is the Rocky Mountain wood tick, *Dermacentor andersoni*. Infestation by a single tick may produce paralysis in sheep, dogs and humans; many ticks are usually required to produce paralysis in cattle, which appear more resistant. Serious livestock losses from tick paralysis have been reported from western Montana, Idaho and Oregon.

There are some 800 known species of ticks in the world, approximately 20 of which are of veterinary importance in the United States. The following is a list of some of the most common species attacking domestic animals.

SOFT TICKS (FAMILY ARGASIDAE)

Fowl Ticks, *Argas persicus, Argas radiatus* and *Argas sanchezi*

These three species of closely related fowl ticks are similar in morphology and biology. *Argas persicus* is found primarily in the northeastern United States while *Argas radiatus* and *Argas sanchezi* are found primarily in the Southeast and Southwest, respectively. Adult ticks (Fig. 9.5) have an oval, flat body, 6-12 mm in length, with a leathery body covering. Color may vary from red to blue-black or gray. All stages attack domestic fowl, especially chickens, where the preferred attachment site is beneath the wings and in other areas where feathers are sparse. Fowl ticks spend most of the time off the host in cracks and crevices of poultry house walls, rafters and floors (Fig. 9.6). They feed primarily at night or in subdued light as poultry roost or nest. Adults may survive 2-3 years without feeding.

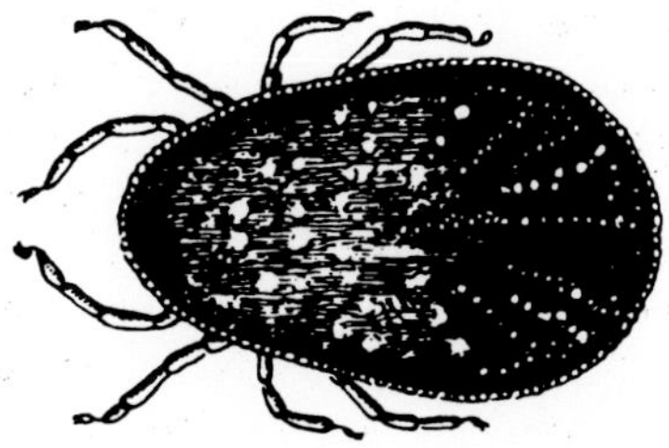

Figure 9.5. Fowl tick. (CDC)

Eggs are usually deposited in poultry house crevices and in ground cracks. After feeding, the adult female deposits eggs in small bunches of 25-245 in seven or more batches; hatching occurs in 8-21 days. The six-legged larva locates a suitable host, feeds for 2-10 days, and then detaches from the host and crawls into a hiding place where it molts in 4-16 days to the first eight-legged nymphal stage. There are generally two nymphal stages but there may sometimes be three; each stage requires a blood meal before molting to the next stage. Nymphs engorge in 30-60 minutes and require 7-28 days to molt to the next stage.

Fowl tick infestation results in an unthrifty flock characterized by dull, ragged plumage, diarrhea and decreased egg production. Infested birds may be weakened from blood loss and irritation and, when parasitized by large numbers of ticks, frequently die. In some areas, including the United States, fowl ticks transmit an organism causing fowl spirochetosis, a fatal disease of chickens, geese, turkeys, guinea fowl and other fowl. Symptoms of the disease include diarrhea followed by loss of appetite, ruffled feathers, pale comb and cessation of perching. Diseased poultry lie down with their heads resting on the ground and usually die during convulsions. This tick also is suspected of causing fowl paralysis and transmitting fowl piroplasmosis.

Ear Tick, *Otobius megnini*

The "spinose" ear tick (Fig. 9.6) obtains its common name from the characteristic spines on the body of the nymphs and from its habit of attaching inside the host's outer ear. Adults have nonfunctional mouthparts and do not feed. The adult's body is devoid of spines and less granular in appearance, with numerous pits on the upper and lower surfaces. The larval stage, which is considerably smaller than the adult, has an oval, striated body and numerous bristle-like, symmetrically arranged hairs. The mouthparts are long and slender. Engorged larvae appear grub-like and are yellowish-white to pink in color. The second-stage nymph is the largest stage and the one most commonly observed. Most records of this species are from cattle, but it is also known to frequently parasitize horses and mules. Other domestic hosts include sheep, goats, hogs, cats, dogs and, on several occasions, has been removed from the ears of man. Among the wild animals that serve as hosts are coyotes, deer, cottontail and jack rabbits.

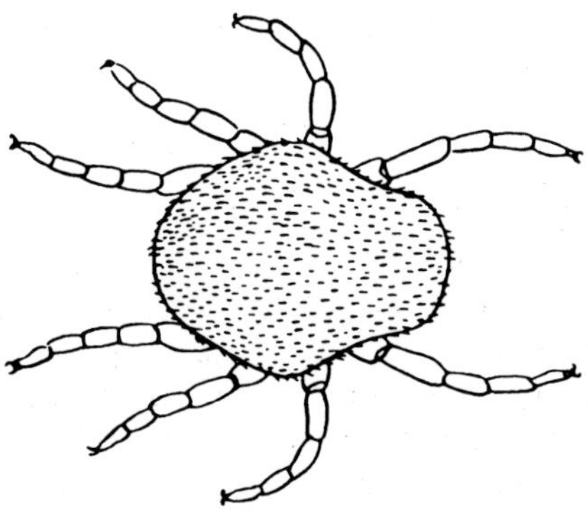

Figure 9.6. Spinose ear tick nymph. (CDC)

Adults do not feed; they are readily found in and around corrals, loafing areas, and especially under and around salt and water troughs in regions where they are abundant. If no males are present, the females remain quiescent up to 18 months; if males are present, females mate and usually begin to deposit eggs in about six weeks. Each female may deposit 350-1,600 eggs during her lifetime, but the average number per female is probably 800-1,000. Egg deposition requires up to six months to complete. Eggs usually hatch in 10-23 days but, under unfavorable conditions of temperature and humidity, this period may be extended. Young larvae crawl up weeds, grass or other objects and, when they contact a host, work their way across the shoulders, neck and head to the ears. Upon entering the ear, they crawl to the inner folds of the outer ear and begin to feed. Larvae do not leave the ear when they complete engorgement, but molt to the nymphal stages within the ear. Larvae require 5-12 days or longer to complete their development. Nymphs attach in the same area as the larvae but occasionally may change their location in the ear. Feeding by the nymphs is usually slow, the two nymphal stages requiring four weeks to seven months. Engorged nymphs detach and drop to the ground where they crawl into cracks, under debris or vegetative material to molt into adults in 7-31 days.

Larvae and nymphs of this species cause serious damage to livestock. Wounds may become infected with pus-forming organisms giving rise to a condition known as "canker ear". Infested animals shake their heads and rub their ears in an attempt to relieve the irritation. The ear canals may become inflamed and frequently the eardrum is affected to such an extent that its function may be impaired. Prolonged and intense involvement may cause deafness. It is not known to transmit any disease organisms.

Pajaroello Tick, *Ornithodoros coriaceus*

This soft tick occurs primarily in California and Mexico, but recently has been discovered in southern Oregon and the northwestern regions of Nevada. It is common in the brushy, foothill areas inhabited by deer and rangeland cattle and horses. Pajaroellos are usually found in deer- or cattle-bedding areas that have coarse soils, fine gravel, or sandy loam mixed with tree litter from nearby scrub oak-chaparral growth. These habitats are common in mountainous areas at elevations from 600-3,000 feet above these elevations to about 8,000 feet, the tick is found under juniper, yellow and pinon pine, and bitterbrush.

Eggs of this tick develop in 10-20 days. Unfed larvae may live up to five months without a blood meal. Feeding time on a host is usually about seven days, after which the larvae drop off and molt in 3-4 weeks to the first nymphal stage. There may be as many as seven stages, each requiring a blood meal lasting 8-100 minutes. Adults feed in 5-50 minutes. Nymphs and adults can survive without feeding for many months. The average life cycle from egg to adult is one year. Females lay 150-300 eggs at a time and produce about four batches a year during a five-year life span.

Cattle and deer are common hosts for this tick. Pajaroellos do not attach themselves to moving hosts; rather, they lie in wait for an animal to recline on the ground and then become active. They have been observed to move slowly toward a host across considerable distances. If not disturbed, they crawl onto the host, feed, and drop back to the ground. Feeding activity can occur either day or night. Nymphal and adult ticks are about 6-12 mm in length, usually about the same color as the soil, and are rarely seen even by an experienced observer. This tick is a vector of epizootic bovine abortion ("foothill abortion") in California and Oregon; the causative agent is a spirochete, *Borrelia coriaceae.*

HARD TICKS (FAMILY IXODIDAE)

Cattle Tick, *Boophilus annulatus*

The cattle tick (Fig. 9.7), also called the Texas fever tick and cattle fever tick, is a one-host tick, with short mouthparts, that lacks contrasting color patterns. All stages attack the same host, most frequently cattle. Larvae prefer the legs, belly and dewlap; nymphs and adults attach on the belly and flanks. Prior to 1889, the cattle tick was widely dispersed by cattle movements throughout the southern states. The species was eradicated from the United States in 1943; however, periodic reintroductions from Mexico occur on cattle and deer despite rigid quarantine measures.

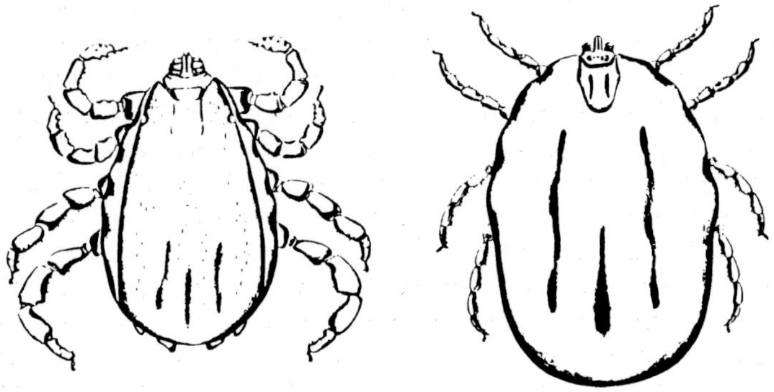

Figure 9.7. Cattle tick male (left) and engorged female (USDA).

Under conditions of summer temperatute and high humidity, the life cycle is short. Eggs hatch in 17-21 days, but may require as long as 5-6 months during cold weather. Since this is a one-host tick and molting from one stage to another occurs on the host, the parasitic period is comparatively short, usually about 3 weeks, but may extend to 59 days. The engorged female usually begins to deposit eggs 34 days after detaching from the host.

In infested areas, the cattle tick may occur on cattle in large numbers and is an important parasite. It is a vector of *Babesia bigemia*, the causal agent of cattle fever or bovine babesiosis; the pathogen is transmitted by the female tick via the eggs to her progency and immature ticks of the next generation infect other hosts.

Southern Cattle Tick, *Boophilus microplus*

This species is similar in appearance, life history and habits to the cattle tick, *Boophilus annulatus*. It is widely distributed throughout tropical and semitropical regions of the world including Central and South America. Sporadic outbreaks occur in south Texas along the Rio Grande River area. This species poses a constant threat to southwestern cattle producers due to its possible introduction on imported cattle or wild animals from Mexico. Within its range, it is an important cattle parasite and vector of both anaplasmosis and babesiosis.

Lone Star Tick, *Amblyomma americanum*

The female lone star tick (Fig.9.8) may be readily identified by a solitary white or gold colored spot on the posterior margin of the scutum. Males are smaller than females and have several scattered white spots on the scutum. Larvae and nymphs have no distinctive markings. This species is distributed from as far west as Texas, north to Missouri, and east in a broad belt to the Atlantic coast. Larvae, nymphs and adults have been recorded from cattle as well as most other domestic animals and many wild hosts. Larvae and nymphs usually parasitize small wild animals and birds; adults usually are found only on large animals. Axillaries, neck, udder or scrotum, escutcheon and tailhead are preferred sites of attachment.

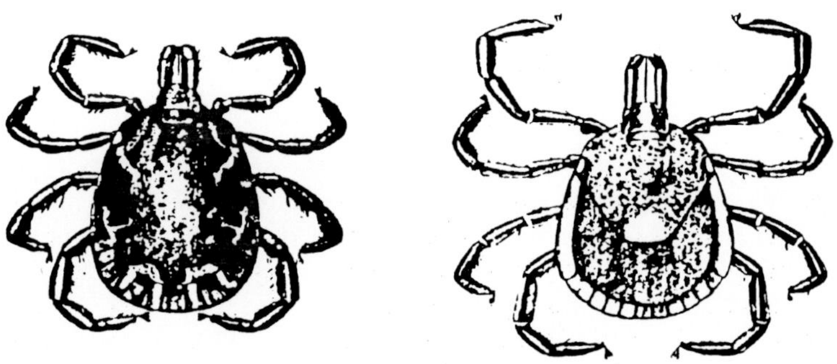

Figure 9.8. Lone star tick male (left) and unengorged female (right). (USDA)

The lone star tick is a three-host tick. Each stage detaches from the host after obtaining a blood meal and molts to the next stage or, in the case of the female, deposits eggs on the ground. The larvae engorge for 3-9 days, drop off and molt 12-30 days later. Nymphs engorge in 3-8 days and molt 13-46 days after detaching from the host. Females feed for 11-24 days, mate, drop off and begin oviposition 5-13 days later. Females deposit an average of 3,000-4,000 eggs which hatch in about 30 days under favorable conditions. The oviposition period averages 21 days. The entire life cycle requires about a year for completion.

The lone star tick is particularly troublesome whenever it attains large populations. The elongate mouthparts result in deep, irritating wounds. Heavily infested animals suffer severely from the bite and may develop an anemic condition due to loss of blood. In addition to its parasitic activity, the species reportedly causes tick paralysis in man and dogs in the eastern and southern states.

Gulf Coast Tick, *Amblyomma maculatum*

The Gulf Coast tick (Fig. 9.9) is somewhat longer and narrower than the lone star tick when viewed from above. The base color of both sexes is chestnut brown, but ornamentation of the scutum differs; the extensive white markings on the female contrasting with the irregularly connected, longitudinal yellowish-white lines on the male. Nymphs and larvae resemble those of the lone star tick. Gulf Coast ticks are established in the United States in regions of high temperature, rainfall and humidity bordering the Atlantic Ocean and Gulf of Mexico, for the most part within 100 miles of the coast; however, populations also occur in Oklahoma. Adults primarily attack large domestic and wild animals such as cattle, sheep, horses, goats, dogs, hogs, wolves, coyotes and others. Larvae and nymphs are found on small wild animals, but principally attack birds, such as meadowlarks and quail, that forage extensively on the ground.

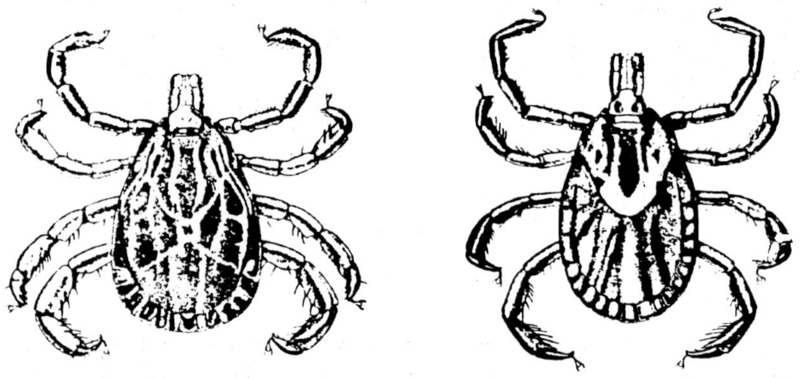

Figure 9.9. Gulf Coast tick male (left) and unengorged female. (USDA)

The Gulf Coast tick is a three-host species. All stages appear very susceptible to desiccation which appears to be an important factor limiting its distribution. Larvae ascend grasses to a height of 4-6 inches off the ground to await a host. On birds, their most important host, they attach to the head and neck, while they are usually found on the head and ears of small mammals. After engorging, usually from 3 1/2-7 days, the larvae detach, crawl into bunches of grasses of weeds and molt to the nymphal stage 9-12 days later. Nymphs are ready to attach, generally about the head and neck of the host, soon after molting; they may be active during the entire year, but are more prevalent during late winter and spring, after engorgement is complete, usually in 4 1/2-7 days, the nymphs detach from the host and seek a molting place similar to that of the larvae; molting time is usually 19-28 days. Temperature appears to be an important factor in stimulating adults to seek a host. Adults are prevalent during the spring, but the majority remain inactive on the base of plants until summer. Females do not completely engorge until fertilized by males. After engorging 14-18 days, the female drops to the ground and seeks a secluded site for oviposition. After a preoviposition period of 2-5 days, the female crawls under debris or makes a shallow excavation in the ground in which to hide. Males usually appear on hosts before females and may remain on the host for several weeks or even months after the engorged females have completed feeding. Unfertilized females have been observed to remain on the host for 70 days. The average number of eggs ranges from 8,000-9,000, with egg-laying activity continuing 13-75 days ot more.

Adult Gulf Coast ticks usually attach inside the outer ear but may be found outside and around the base of the ear. The elongate mouthparts produce deep, irritating wounds which cause animals to rub their head and ears; bites result in severe inflammation, swelling, cracking and scabbing. Severely damaged ears may curl inward resulting in a characteristic appearance termed "gotch ear." A rickettsia from Gulf Coast ticks which had fed on sick sheep experimentally produced a condition termed "maculatum disease."

Cayenne Tick, *Amblyomma cajennense*

The "cayenne" tick (Fig. 9.10) is a medium-sized tick with characteristic ornamentation on the scutum formed by reddish-brown stripes on a pale background. It is a three-host tick. In the United States, this species is limited to southern Texas; however, it is widespread throughout Mexico and Central and South America. Adults most commonly parasitize horses and prefer to attach between the legs or on the abdomen, but all stages are frequently found inside the ears as well as on the flanks, withers, mane and tail. On cattle, the tick may be found attached to any part of the body. Little is known about the life cycle of this species; it appears similar to that of the lone star tick.

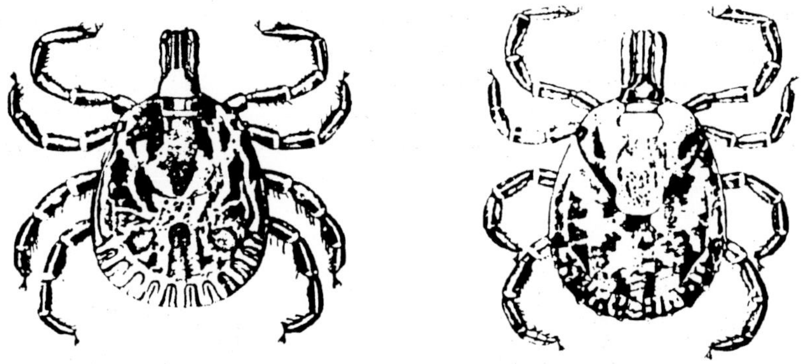

Figure 9.10. Cayenne tick male (left) and unengorged female (right). (USDA)

Rocky Mountain Wood Tick, *Dermacentor andersoni*

This species of three-host tick occurs throughout the western one-third of the United States; development is favored by moisture and protection from direct sunlight in brushy areas. Larvae and nymphs parasitize a wide variety of wild rodents while adults most frequently occur on large domestic animals such as horses and cattle where they attach principally between the legs and on the belly and dewlap.

Adult ticks appear in the spring and climb stems and tips of low vegetation present along animal trails in pastures or in forested areas to wait for a passing animal. Once attached to a host, mating occurs during the process of taking a blood meal. After feeding for approximately one week, the female detaches from the host and deposits 5.000-6,000 eggs over the next 30 days. Hatching occurs in 4-6 weeks; seed ticks climb on grass or other plants and attach to small rodents that pass by. Engorgement is completed in about one week, after which the larvae drop off and molt in 3-4 weeks to the nymphal stage. Nymphs seek shelter from the hot summer temperatures and become inactive until the following spring. The following year, they attach to a second host, feed for about one week, drop off, and remain on the ground for 1-2 months before molting into the adult stage during late summer or early fall. Warm weather again causes the adults to seek shelter from which they reemerge the following spring. Thus, two years are usually required to complete the life cycle although this interval may extend to 3-4 years depending on weather conditions and the availability of hosts. Larvae which cannot find a host usually die in less than three months, whereas nymphs and adults are more hardy and may survive without feeding for nearly a year. There is an overlap of spring emergence by nymphs and adults, the former to feed on small rodents and the latter to engorge on larger domestic and wild animals; with the advent of hot weather, seed ticks begin to appear on small rodents.

Dermacentor andersoni has been incriminated as a vector of bovine anaplasmosis and canine babesiosis. Toxins injected during feeding cause tick paralysis in livestock, wild mammals and man. Hides punctured by numerous tick bites reduce leather quality. This species is the major vector of Rocky Mountain spotted fever to man in the western United States.

Pacific Coast Tick, *Dermacentor occidentalis*

The Pacific Coast tick occurs in the narrow land strip between the Sierra Nevada Mountains and the Pacific Ocean from Oregon to southern California. The life cycle and behavior of this three-host species are similar to those of *D. andersoni*, to which it is closely related, with the various stages using small rodent hosts and larger domestic and wild mammals to complete a 2-3 year developmental period. Larvae and nymphs normally feed on various rodents and other small mammals; adults typically attack deer, cattle and horses. It causes tick paralysis in cattle, horses and deer, and transmits bovine anaplasmosis.

American Dog Tick, *Dermacentor variabilis*

This species (Fig. 9.11) is a three-host tick that occurs on hosts throughout the year in the southern part of its range, but is more numerous in the spring. It is widely distributed over the eastern two-thirds of the United States; high humidity favors development so that populations are more abundant in coastal areas. The life history is similar to that of *Dermacentor andersoni*. Dogs are the preferred host of the adult, but it has been collected from numerous wild and domestic animals as well as from man. Larvae and nymphs prefer small rodents. This species transmits Rocky Mountain spotted fever to man and dogs in the eastern United States and is also a vector of canine babesiosis.

Figure 9.11. American dog tick male (left) and unengorged female (right). (USDA)

Winter Tick, *Dermacentor albipictus*

The scutum of the winter tick (Fig 9.12) has extensive white markings or is mostly white streaked with brown. This tick is widely but unevenly distributed across the northern tier of states from Maine to Oregon and throughout the western states as far south as Texas. It is abundant in the woods and shrubs of upland and mountainous areas where it is commonly found on large domestic and wild animals. Horses and cattle are the most commonly parasitized domestic animals; wild hosts include deer, elk, antelope, and moose.

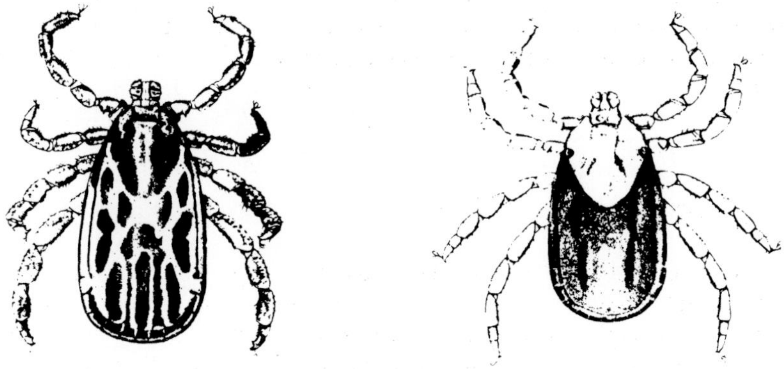

Figure 9.12. Winter tick male (left) and unengorged female (right). (USDA)

The winter tick is a one-host tick. It attaches to a host as the larva or "seed tick" and remains on the same host to molt from the larval to the nymphal stage and from the nymphal to the adult stage. The adult female detaches after engorgement and fertilization to oviposit on the ground. As the common name implies, this tick is usually found on animals during the winter. During the summer, larvae remain quiescent around the base of clumps of grass or other plants and do not react to a host even if one grazes nearby. During the shortened day lengths and cool nights of September or October, larvae actively seek hosts; the adult population reaches its peak during the latter part of November or December. In January, the population begins to decline and, by February and March, only a few adults are found on animals. The total parasitic period is usually 5-7 weeks but may be as long as 18 weeks. Each female usually deposits 3,000-4,000 eggs that hatch in 3-6 weeks.

Since this species is most abundant during the winter, infestations in the long winter coat of the host may be frequently overlooked. These ticks cause loss of appetite and general debilitation. Blood loss from heavy infestations weakens and may kill horses. Heavy infestations also may result in an edematous condition known as "water belly". Winter ticks also have been found to be a vector of bovine anaplasmosis under experimental conditions.

Tropical Horse Tick, *Dermacentor nitens*

The tropical horse tick (Fig. 9.13) occurs in south Texas, Florida, Mexico, and Central and South America. This species is a one-host tick which usually attaches inside the ears of horses, mules and donkeys. Ticks also may be found in the nasal diverticula and mane in heavy infestations. Blood engorged females detach from the host and commence oviposition in 3-15 days. The entire cycle may be completed in as few as 27 days. This tick causes intense irritation and soreness of the ears and is known to transmit equine babesiosis, a serious disease of horses.

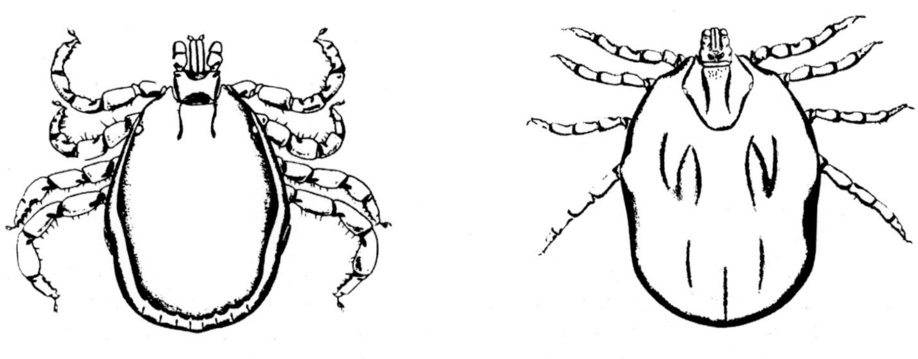

Figure 9.13. Tropical horse tick male (left) and engorged female (right). (USDA)

Blacklegged Tick, *Ixodes scapularis*

The blacklegged tick (Fig. 9.14) is a three-host species common in the eastern, central and southern United States. Adults prefer to feed on larger mammals, such as cattle. horses, deer and dogs where they usually attach to the head and neck. Larvae and nymphs feed primarily on birds, small mammals and reptiles. Adults are more abundant on hosts from late fall to spring; immature states are more abundant during the spring and summer. The life history is similar to that of the lone star tick. High relative humidity and moderate temperatures are required for development. Under natural conditions, one generation a year is probably normal. Considerable tissue damage to the host may result due to the long mouthparts which are often broken off and remain embedded in the skin when ticks are forcibly removed. Transstadial transmission of bovine anaplasmosis has been reported and ticks have been found naturally infected with the causative agent of tularemia. More recently, this species is also suspected to be a vector of Lyme disease to man and some species of animals.

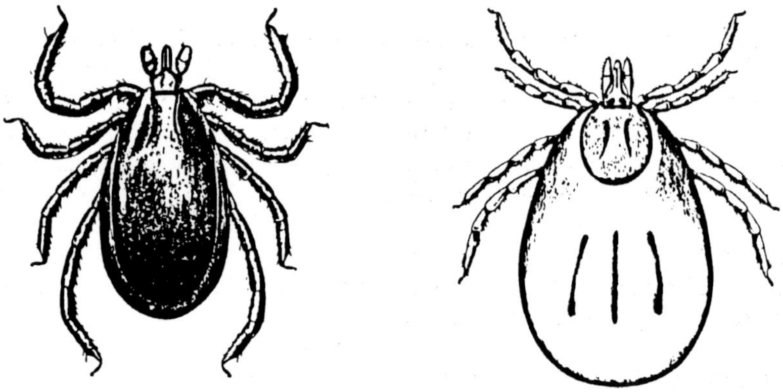

Figure 9.14. Blacklegged tick male (left) and engorged female (right). (USDA)

Western Blacklegged Tick, *Ixodes pacificus*

This species occurs primarily along the Pacific Coast of the United States. The life history is essentially similar to that of *Ixodes scapularis*. The life cycle may be completed in less than one year in the coastal regions of central and northern California where moderate temperatures and a high level of moisture is common. The immature stages occur commonly on lizards, ground inhabiting birds and, less frequently, on small rodents and rabbits. Adults are found primarily on large mammals; dogs are frequent hosts. This species is the principal vector of Lyme disease on the west coast.

Northern Deer Tick, *Ixodes dammini*

Northern deer ticks occur primarily in the northeastern United States. The life cycle of this three-host tick apparently spans two or more years. Larvae most frequently engorge during July through September, detach from the host, and molt at a time permitting resultant nymphs to feed during May through July of the following year. Adults are active mainly during the fall. Larvae and nymphs are parasitic primarily on the white-footed mouse although other small rodents, rabbits and deer are occasionally attacked. Adults are parasitic almost exclusively on white-tailed deer. This species is the principal vector of Lyme disease, an arthritic condition caused by the spirochete, *Borrelia burgdorferi*, in the northeastern United States.

Brown Dog Tick, *Rhipicephalus sanguineus*

The brown dog tick (Fig. 9.15) is believed to be a native African species whose close association with domestic dogs has facilitated its spread throughout the warmer regions of the world. The brown dog tick is probably the most widely distributed tick species in the world. It is common throughout the United States where it is parasitic almost exclusively on dogs. Larvae and nymphs attach almost anywhere on the host. while adults are frequently found on the neck and head.

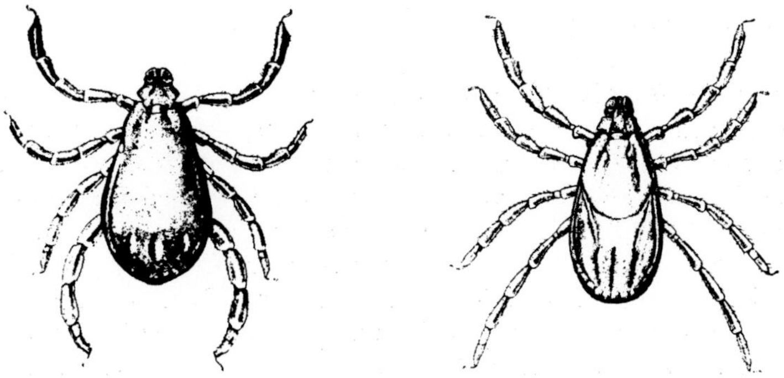

Figure 9.15. Brown dog tick male (left) and unengorged female (right). (USDA)

The brown dog tick is a three-host tick. Oviposition begins on the third day after females drop off the host, and within 21-29 days, 4,000-5,000 eggs may be deposited. Females oviposit under stones or straw, or between boards, under plaster, or in crevices in walls as high as 15 feet above the ground. Eggs hatch in about 19 days. Larvae typically congregate near the bottom of walls waiting for a suitable host. Once attached, larvae may engorge in three days; molting may occur in a minimum of six days. Nymphs engorge in four days and molt in 12. Under favorable conditions, the life cycle may be complete in 63 days so multiple generations per year may occur.

The brown dog tick is an exceedingly troublesome pest of dogs. It is a vector of both canine babesiosis and canine ehrlichiosis (malignant canine rickettsiosis), and is also believed to transmit *Salmonella enteritidis*, which causes a paratyphoid disease in dogs and laboratory animals.

Rabbit Tick, *Haemaphysalis leporispalustris*

Rabbit ticks (Figure 9.16) are easily recognized by their comparatively small size and an angular projection of the second segment of the palps. This species is widely distributed throughout the entire United States. Rabbits are preferred hosts, although dogs and cats may be attacked in rural areas. Adults usually attach to the ears, around the eyes and about the head of their host.

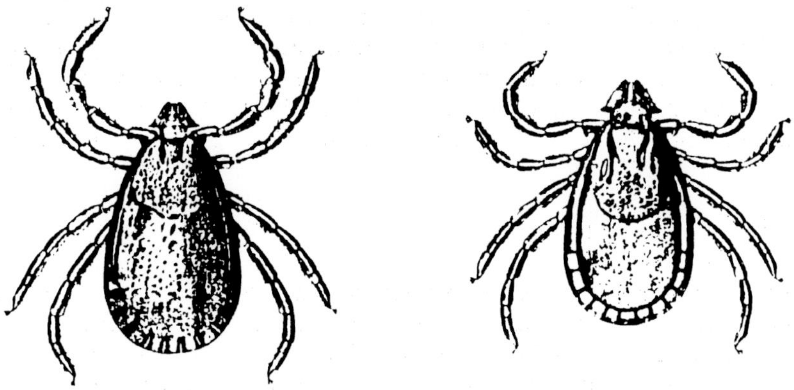

Figure 9.16. Rabbit tick male (left) and unengorged female (right). (USDA)

The rabbit tick is a three-host tick; all stages are active throughout the year in warm climates. This species has apparently adapted its life cycle to the nocturnal habits of its host. Engorged females, which complete feeding in about 17 days, do not detach from the host during its nighttime wandering, but drop off only during the day when the host is inactive in its resting site. Fifty-nine to 2,400 eggs are deposited. Eggs hatch in a minimum of 22 days and the larvae have ready access to the host. Larvae engorge in 4-10 days and detach; molting occurs in 18-134 days. Nymphs attach to the same or a different host, complete feeding in 4-8 days, detach and molt in 13-124 days to the adult stage. This species does not ordinarily attack domestic animals; however, gross infestations may weaken and kill rabbits and birds.

CONTROL

Since fowl ticks generally feed on hosts for only a relatively short period of time, premise treatment with an approved insecticide is usually the most effective control measure. Areas where these ticks hide, such as cracks and crevices of poultry house walls and floors, should receive special emphasis. Removal of birds from infested premises is usually not practical since this species may survive in an empty house as long as two years without feeding. Modern poultry house construction, which minimizes hiding places, have greatly reduced economic losses associated with the fowl tick. Spinose ear ticks may be controlled by application of various insecticide formulations directly into the auditory canal. The Pajaroello tick is difficult to control due to its widespread habitat.

Hard tick control, on the other hand, is usually directed against the infested host to which these ticks are attached for several days. Since the goal of successful tick control is the prevention of the development of engorged female ticks, which may deposit large numbers of eggs to propagate the next generation, treatment intervals for one-host ticks, which spend a longer time in association with the host, are longer than those for three-host ticks. Method and frequency of treatment will vary with the species and the attachment site on the host; when planning control strategy, ticks are generally regarded as "ear ticks" or "body ticks". Insecticide formulations, such as liquids, dusts or salves, should be applied directly in the ear of the host to control those species characteristically attaching in this area; insecticide-impregnated cattle ear tags may also be effective for ear tick control. Infestations of body ticks necessitate thorough treatment of the entire skin surface of the host. Dipping is the preferred method of application when eradication is the goal since it promotes uniform host coverage; however, it is not economically feasible for most livestock producers. Pasture vacation or rotation, seasonal burning and brush clearing aid in the control of those stages not attached to a host.

CHAPTER 10

MITES

CLASSIFICATION

As discussed in Chapter 9, mites and ticks belong to order Acari, class Arachnida. However, while all ticks are placed in the single suborder Metastigmata, parasitic mites of domestic animals usually are representatives of three different suborders. Characters associated with these suborders and representative species of each are as follows.

Suborder Mesostigmata: These mites have a single pair of spiracles lateral to the legs, usually associated with an elongated peritreme; if there is no peritreme, the mite is usually a highly specialized parasite of the respiratory tract of vertebrates. Haller's organ is absent; the hypostome is neither toothed nor developed for piercing. Representatives include such species as the chicken mite (*Dermanyssus gallinae*), northern fowl mite (*Ornithonyssus sylviarum*) and tropical fowl mite (*Ornithonyssus bursa*).

Suborder Prostigmata: Mites of this suborder have a pair of spiracles on or near the head region (the gnathosoma). The chelicerae are usually modified for piercing; palps are usually free and highly developed. Haller's organ is absent. The coxae seldom form conspicuous internal projections beneath skin. There are no anal suckers. Representatives include chiggers (*Trombicula* and allies in the Family Trombiculidae) and follicle mites (*Demodex* species).

Suborder Astigmata: These mites have no spiracles; a few have a system of tracheae that open through stigmata and porous areas on various parts of the body. There is no Haller's organ. The coxae form conspicuous internal projections beneath the skin of the venter of the body. The mouthparts are usually adapted for chewing; they have strong chelae. The palps are simple. Anal suckers are often present. Representatives include the mange and itch mites (*Sarcoptes, Psoroptes, Chorioptes* and *Notoedres*).

MORPHOLOGY

Mites differ so greatly from one species to another that a short, generalized description cannot be made; however, the following brief description of a typical female rat mite, *Laelaps echidnina* (Fig. 10.1), will establish some familiarity with the basic morphology of these arthropods. The mite body is fused into a single region with no well-defined segmentation. The "head" region is termed the capitulum or gnathosoma. Replacing the mandibles and maxillae of true insects, mites have cutting structures, termed chelicerae, with two scissor-like blades called chelae. The structure of the chelicerae is useful in identification. The chelae may have teeth, they may be smooth resembling the blades of scissors or they may be drawn out to a needle-like point. Palps may be simple or greatly modified by specialized hairs. The tritosternum, present only in the Mesostigma, is a forked structure located posterior to the capitulum.

Adult mites possess four pairs of legs, which differ from those of insects in having one additional segment, the patella, and, in the majority of species, an unsegmented tarsus. In many mites, the tarsus bears two claws and a fleshy pad, the caruncle. The bases of the legs, or coxae, are inserted into cavities close to one another in the suborder Mesostigmata, which includes some of the rodent and bird mites. However in the other suborders, the first and second coxae are situated close together but are widely separated from the third and fourth coxae as in the scabies and mange mites.

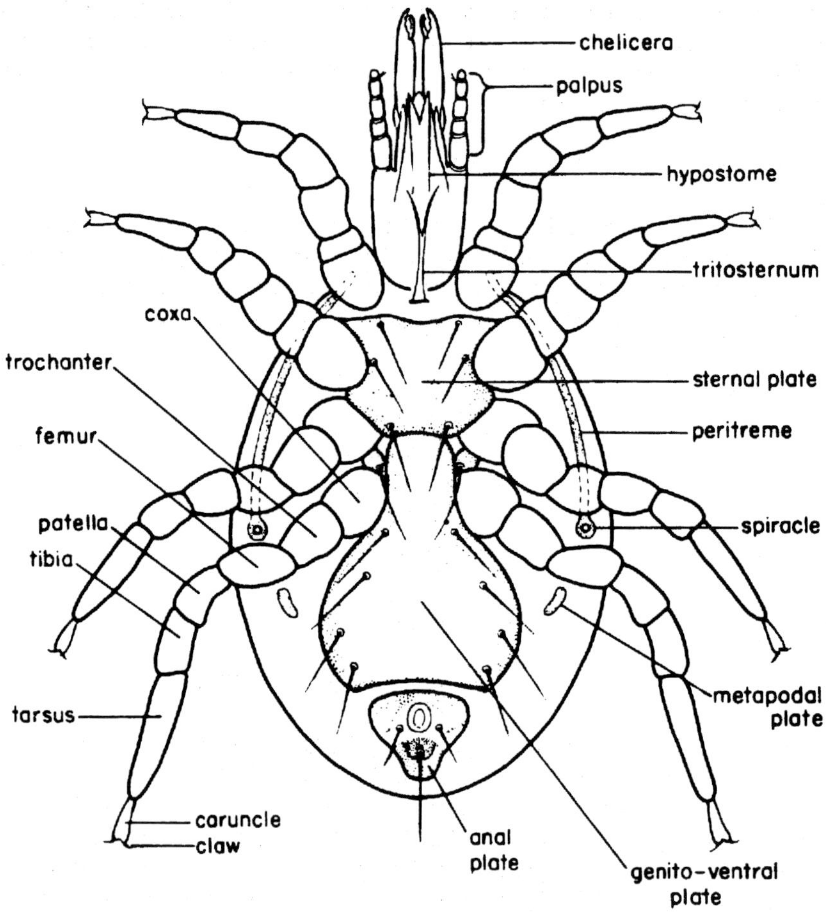

Figure 10.1. Diagram of female rat mite, *Laelaps echinina*, with parts labeled. (CDC)

The ventral side of many mites bears three plates (the sternal plate, the genito-ventral plate and the anal plate), all of which are important in classification because of their various shapes and the variable numbers of hairs located on them; metapodal plates may be present behind the fourth coxae. Some species have a single dorsal plate while other species and many protonymphs have two or more dorsal plates. Many mites breathe directly through the cuticle; others breathe through spiracles, which are the terminal points of internal tubes (peritremes) of variable length. The genital opening is at the anterior end of the male, often between the second pair of legs.

GENERALIZED LIFE CYCLE

The generalized life cycle of mites is similar to that of ticks except that more than one nymphal stage is usually present. Typically, female mites deposit eggs which hatch into larvae that pass through two or more nymphal stages to eventually mature into adults. Larvae have only three pairs of legs and may be distinguished by this character from later stages, all of which have four pairs of legs. The first nymphal stage (protonymph) has four pairs of legs, one or more plates on the dorsal side and a single, short sternal plate with three pairs of setae. The second nymphal stage (deutonymph) has four pairs of legs, a single dorsal plate and a longer sternal plate, extending at least to the fourth pair of coxae, that bears four pairs of setae.

POULTRY MITES

Several species of mites attack domestic fowl. Stages in the life cycle include: adult males and females, eggs, six-legged larvae that may or may not feed and usually two eight-legged nymphal stages, with feeding in both, or in some cases only one, stage. Poultry mites usually occur on or under the skin or feathers; however, some species may be found in body tissues, feather quills or nasal and respiratory passages such as the air sac. They live on bits of skin or feather or by piercing the skin or tissue and sucking blood or body fluids. Some species transmit serious and often fatal, disease-producing organisms.

Northern fowl mite, *Ornithonyssus sylviarum*

The northern fowl mite (Figs. 10.2) is the most important and common ectoparasite of poultry in the United States. The host range includes all domestic fowl and many species of wild birds; it is distributed throughout the temperate regions of the world. Adults are about 1 mm long with an oval to elongate body shape; coloration is usually dark red to black, although variations sometimes result in a banded appearance.

Figure 10.2. Northern fowl mite. (TAEX)

The entire life cycle (Fig. 10.3) is usually passed on the host. Adult females normally take a blood meal and deposit 2-5 eggs in the fluff of feathers with two days. Eggs hatch in 1-2 days into a non-feeding larval stage that lasts less than a day. The first nymphal stage (protonymph) feeds twice during its 1-2 day developmental period; the second nymphal stage (deutonymph) does not feed and completes development to the adult in less than one day. The complete cycle may be as short as 5-7 days. Although large numbers of eggs are not deposited, populations can increase rapidly after initial infestation. Under optimal conditions, newly infested birds may support in excess of 20,000 mites within 9-10 weeks. Adults spend most of their lives on the host, but during heavy infestations may be found on eggs, egg flats, cages and poultry house walls; individuals may survive off the host up to two months at room temperature.

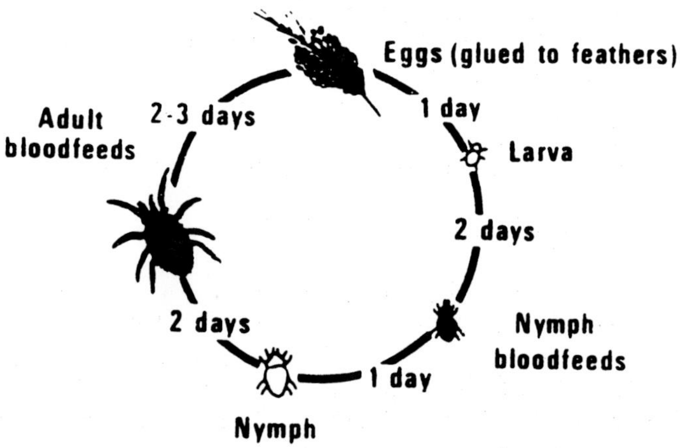

Figure 10.3. Northern fowl mite life cycle. (CCEX)

Mites congregate near the vent, tail, back and neck, sucking blood and irritating the host. Infestations of large numbers of mites can reduce egg production by 10-20%; small to moderate numbers do not affect egg production, although egg handlers may complain of bites and irritation. Stressed birds usually reduce food intake, lose weight rapidly and exhibit pale combs due to anemia. Feathers become soiled from the numerous eggs, cast skins and excrement of the mites. Scabs often form on the skin of infested fowl. Fowl mites are most abundant during the winter.

Flocks should be examined frequently to monitor infestations; darkened feathers in the vent area are indicative of high mite numbers. To be effective, chemical control should be carried out with spraying equipment capable of sufficient pressure to penetrate the feathers; it is very important to direct applications to the vent area for effective results. Roosters should be treated carefully and thoroughly to avoid reinfestation in breeding stocks. Heavily infested premises may require treatment as well. Infestations should be prevented to the greatest extent possible by proper cultural practices. New birds introduced into a house should be mite free; equipment may require steam cleaning since mites may live for several months off the host. Wild birds should be excluded to the extent possible.

Tropical fowl mite, *Ornithonyssus bursa*

This species is easily confused with the northern fowl mite, *Ornithonyssus sylviarum*, to which it is closely related. Populations tend to congregate in the downy feathers in the vent area. The life cycle, although not well known, is passed on the host. Eggs hatch in approximately three days and the nonfeeding larval stage develops in about 17 hours. The first nymphal stage develops in 1-2 days, but the development period for the second nymphal stage is unknown. Adults probably feed intermittently since they spend considerable time on the host. They normally are found only on the host or in nests and can live only about ten days off the host. This mite is seldom reported from caged poultry in the United States at the present time.

Chicken mite, *Dermanyssus gallinae*

The chicken mite, also known as the red mite, roost mite or poultry mite, occurs in tropical and temperate zones where domestic fowl or caged birds are kept. All kinds of domestic poultry and many species of wild birds are attacked. Adults are slightly smaller then northern fowl mites (0.75 mm long); coloration is grayish, but when engorged with blood, it is red to nearly black. When not feeding, chicken mites may be found in secluded cracks and crevices of roosts, walls, ceilings and floors of the poultry house, especially during the day (Fig. 10.4). At night, they crawl onto the host, pierce the skin and suck blood. This species is seldom found in caged layer operations since there are few places to hide and develop under such conditions.

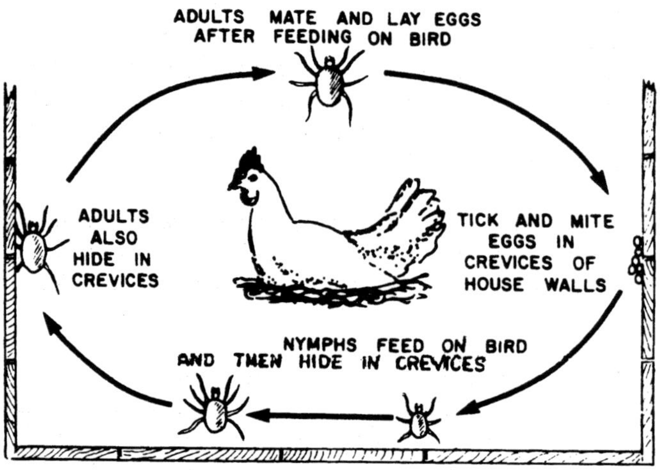

Figure 10.4. Chicken mite life cycle. (CCEX)

Females usually begin oviposition one day after feeding; eggs are deposited in several batches of 25-50 in cracks and crevices. Hatching occurs in 2-3 days, and the non-feeding larvae develop in 1-2 days. The first and second nymphal stages, lasting 1-2 days each, feed soon after molting. Under favorable conditions, the life cycle can be completed in about ten days so there are numerous generations per year. Mites are inactive during the winter except in heated buildings.

Infestations may go unnoticed unless birds are examined at night. Heavy populations debilitate birds resulting in lowered egg production and decreased weight gains; death has been reported in young birds. This mite is also a vector of fowl spirochetosis, a fatal disease of chickens, geese, turkeys and guinea fowl. Control of chicken mites is essentially directed toward sanitation and treatment of the premises with approved acaricidal solutions.

Scaly-leg mite, *Knemidocoptes mutans*

Scaly-leg mites (Fig. 10.5) are widely distributed throughout the world and attack chickens, turkeys, pheasants and many species of passerine birds. Researchers suspect that wild birds transmit the mites to domestic flocks. Females are small (0.25-0.5 mm), oval mites with short, stubby legs.

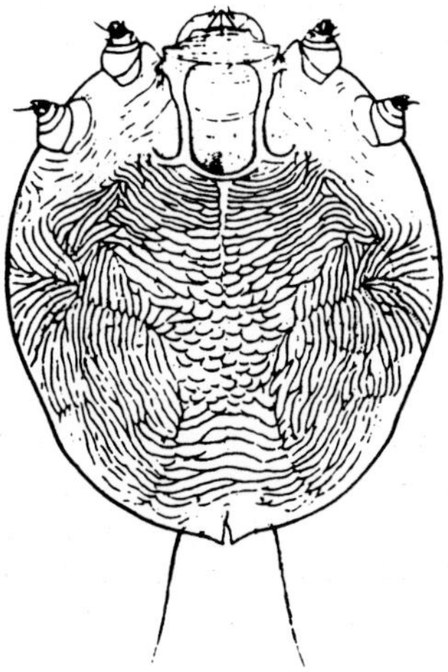

Figure 10.5. Scaly-leg mite. (CCEX)

The scaly-leg mite is a permanent parasite, passing the entire life cycle in the skin of the legs. Females burrow under scales on the feet and legs of poultry where they deposit up to 20 eggs during a life span of two weeks. Eggs hatch in 4-8 days into six-legged larvae that soon molt into nymphs. Nymphs develop into mature males and immature females within one week. The immature female transforms into a mature egg-laying female shortly after fertilization. The entire life cycle probably requires 10-14 days.

Burrowing beneath scales on feet and legs results in localized tissue irritation, cellular proliferation and the gradual formation of crust-like lesions or scabs. If untreated legs become swollen and heavily encrusted, deformity of the shank may result. Death may occur in severely affected birds because they lose the mobility needed to take food and water.

Identification of scaly-leg mite infestations is not easy. Suspect lesions should be covered with an application of warm vegetable oil to loosen the scabs which can then be scraped for microscopic examination. Differential diagnosis is important because similar lesions may result from other conditions such as leg wounds or poor nutrition. Legs of infested birds should be dipped in a solution of an approved acaricide. Retreatment in approximately eight days is necessary to kill mites which have hatched from eggs remaining in tunnels in the skin.

Depluming mite, *Knemidocoptes gallinae*

Depluming mites occur throughout the United States; hosts include pigeons, pheasants, geese, canaries, chickens and many species of wild birds. This mite is essentially similar to the scaly-leg mite, but is smaller and more oval.

Developmental stages include egg, larva, nymph, adult male and immature and mature female. Little is known about the life history of this species, but transformation from immature to mature female is believed to occur after fertilization. The fertilized female begins depositing eggs within a few hours after burrowing beneath the skin and continues ovipositing at 2-3 day intervals for approximately two months. Eggs hatch in about five days. The complete life cycle requires 10-14 days.

The depluming mite burrows into the skin at the base of the feathers, on the back, top of the wings, around the vent and on the breast and thighs. It causes intensive itching. often resulting in feather pulling. Fowl may lose feathers over large areas of the body. Infestations, especially noticeable during the spring and summer, may spontaneously disappear in the fall. Treatment consists of dipping infested birds in a solution of an approved acaricide.

Turkey Chigger, *Neoschoengastia americana*

Several species of chiggers, most of which belong to the genus *Trombicula*, parasitize wild and domestic animals; however, the most economically importance species in the United States is the turkey chigger (Fig. 10.6). This species extends across the southern states from the Carolinas to California. It is particularly abundant in areas with hard soils that crack open during hot, dry summers and in areas where rock outcroppings occur. Populations begin to increase in late April or May, peak in June and decline during late July or August. There may be an increase in September or October, but by late October or November, it disappears from the host. Domestic hosts are turkeys and chickens; wild hosts include quail and other species of wild birds.

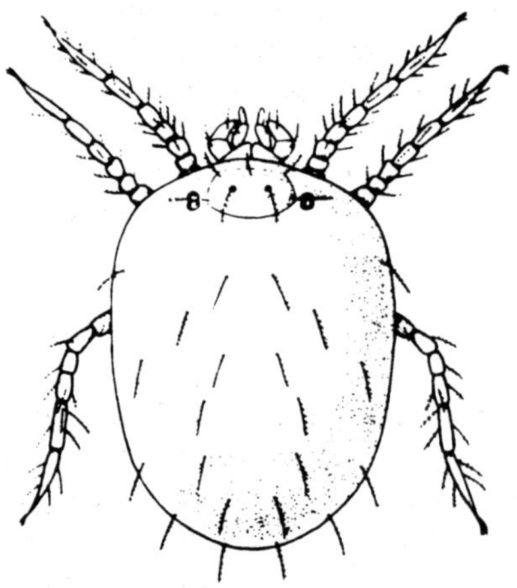

Figure 10.6. Turkey chigger larva. (TAEX)

The life cycle of the turkey chigger is considerably different from that of other parasitic mites of poultry. Eggs usually are deposited singly in the soil. After a 4-6-day incubation period, eggs hatch into a deutovum, which remains in the egg shell fragments for about one week before the six-legged larva emerges. Larvae crawl around rapidly and may survive up to two weeks or more in search of a host. The larval stage (Fig. 10.6) is the only parasitic stage in the chigger life cycle. Larvae normally do not burrow into the skin or suck blood; rather, they inject a digestive enzyme into the wound that liquifies host tissue which is subsequently sucked up, leaving a tube termed a stylostome. The digestive enzyme that hydrolyzes the host's tissues is probably responsible for the localized reaction that results in the characteristic "chigger bite". Engorgement usually requires 1-4 days but in some instances may take up to a month. When feeding is complete, larvae drop to the ground, burrow into upper layers of the soil and become quiescent. Within the larval skin, a nymphochrysalis develops and, about one week later, the eight-legged nymph emerges. The nymph is larger, has a figure eight-shaped body and is more hairy than the larvae. Nymphs are probably predaceous on insect eggs and other small arthropods. Within a week, nymphs enter a quiescent stage, the imagochrysalis, and emerge as adults in another week. Adults are larger, hairier and sexually mature. The sexes are similar in size and appearance but differ in appearance of the genital opening. They have essentially the same feeding habits as do the nymphs. Adults commence oviposition within one week and continue egg laying as long as favorable conditions exist. Under laboratory conditions, observers have counted as many as 4,764 eggs deposited from a single female within 23 days.

Turkey chiggers feed in clusters on the thighs, breasts, underside of wings and around the vent area resulting in scabby lesions that require about three weeks to heal after the engorged larvae detach from the host. These lesions and scabs result in downgrading of carcasses, the major economic loss associated with this species. Control consists of treatment of the soil where turkeys range with an approved acaricide.

LIVESTOCK MITES

Most species of parasitic mites of livestock (cattle, horses, swine, sheep and goats) in the United States belong to four groups (genera): *Psoroptes, Sarcoptes, Choriptes* and *Demodex.* The first three groups are commonly referred to as scab, mange or itch mites; the last group consists of the follicle mites. Authorities disagree as to whether distinct species of these four genera exist as parasites of different animal hosts or whether they are strains or varieties of species which have become adapted to specific hosts. Regardless of their taxonomic status, however, the life history, nature of damage and method of control is similar for all members of the same genus. Cattle and horses are attacked by all four genera; sheep and goats are parasitized by *Psoroptes, Sarcoptes* and *Chorioptes*; while only *Sarcoptes* and *Demodex* occur on swine. Infestations of scab or mange mites result in a condition termed scabies, characterized by thick, crusty lesions of the skin and considerable loss of hair. Lesions produced by follicle mites typically appear as small nodules; there is usually no change in the hair coat.

Differential diagnosis is important because lesions produced by these various species of mites may appear clinically similar and may even be confused with infestations of lice during the early stages; sarcoptic mange does not generally respond to treatment as favorably as the other types. Positive identification depends upon microscopic identification of the mite following skin scrapings or the removal of small scabs. Certain species of scab or mange mites attacking cattle, sheep and goats are subject to quarantine regulations, and diagnosed infestations must be treated with an approved acaricidal solution under the supervision of appropriate authorities. Follow-up treatments are typically necessary to kill those stages which have hatched from eggs, not affected by the initial treatment, before they have the opportunity to mature, mate, and lay additional eggs.

Psoroptic Scab Mites, *Psoroptes*

Psoroptic scab or mange mites (Fig. 10.7) are small (0.5 mm long), oval mites with tapering mouthparts that are longer than broad. The four pairs of legs in the adult mite are rather long; the first two pairs are close together and widely separated from the third and fourth pair which also are very close together. Cup-like tarsal suckers are located on segmented, stalk-like structures termed pedicels at the ends of the first, second and fourth pairs of legs of females; the third pair of legs, devoid of tarsal suckers, terminates in a single, long seta or hair. Males have tarsal suckers on the first three pairs of legs; the fourth pair is somewhat reduced.

Although *Psoroptes* mites have been reported from cattle, sheep and goats, and horses; infestations are of economic importance primarily to cattle on rangeland in the western states and to the cattle feedlot industries in Texas, Oklahoma, Kansas, Nebraska and Colorado. The disease has been eradicated from sheep in the United States in 1973, however, there is constant danger of its reestablishment. Mites spread from infested to non-infested animals primarily by direct host contact. Frequently, infested bulls transfer the mites to large numbers of cows during mating. Rubbing objects used by infested animals may also constitute a source of infestation. Research indicates that mites may survive up to ten days off the host; experimental infections from contaminated pens have not succeeded after infested animals have been removed nine or more days.

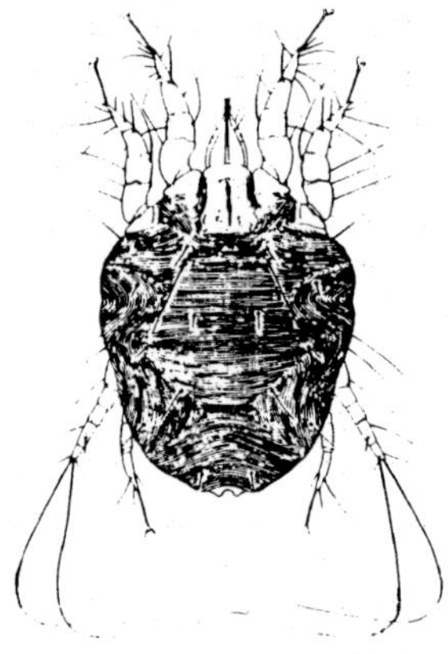

Figure 10.7. Psoroptic scab mite. (USDA)

Psoroptic scab mites complete their entire life cycle on the host. Each female deposits approximately 15-24 eggs on the skin or base of a hair. Eggs deposited on or adjacent to the skin hatch in 3-4 days, however, those separated from the skin by scabs or other debris may require up to five days to hatch. Eggs removed from the host may hatch in ten days but the larvae are unable to survive. The four-legged nymphs are of two sizes. Smaller nymphs feed and molt into sexually mature males in 3-5 days, whereas larger ones feed and molt to the pubescent or copulatory female in 3-4 days. Females usually appear about 1/2-1 day before the males. Adults mate in two days and the pubescent or copulatory female molts again into an egg-laying female which begins oviposition about one day later.

Figure 10.8. Psoroptic scab mite infestation. (B. C. Clymer)

Although most of the host's body that is thickly covered with hair may eventually be infested (Fig. 10.8), lesions usually appear first on the withers, over the back or around the tail head. The disease is seldom detected at this early stage however. There is considerable itching and irritation as mites become more numerous, and infested animals lick themselves and rub against objects, possibly developing abcesses on the shoulders and rump areas. Papules typically form and rupture so that serous exudates flow onto the skin surface and become mixed with dirt and contaminated with bacteria. In the early stages, these scabs may be about the size of a pea; however as the mites move to the periphery of the lesions in search of healthy tissue, lesions increase in size and eventually coalesce. As severity increases, the scabs become thick and crusty and the skin thickens. Irritation at this stage is extremely intense, frequently accompanied by emaciation and anemia.

Mites multiply rapidly and produce the most severe skin lesions during the fall and winter. Lesions tend to disappear during the summer, but sufficient numbers of mites survive to repopulate the host during the onset of cool, fall weather. If untreated, cattle scabies becomes a serious disease affecting animals of all ages, classes and conditions. The disease develops more rapidly and is more severe in weak, malnourished animals; it may then spread rapidly to other members of the herd, especially during the winter. Lesions may be visible 15-45 days after exposure but in some cases the interval may be much longer. Typical symptoms are weight loss or reduced weight gains, reduced milk flow, unthriftiness and even death. Severity of the disease is compounded by cold, rainy weather coupled with a poor state of nutrition.

Sarcoptic Itch Mites, *Sarcoptes*

Sarcoptic itch or mange mites (Fig. 10.9), also referred to as true scab mites and barn itch mites, generally resemble psoroptic scab mites (*Psoroptes*) but are slightly smaller with a rounded rather than oval shaped body. Under magnification, the dorsal surface of the body appears wrinkled with transverse lines; a number of strong backward-projecting, lance-like spines are evident on the posterior portion of the body. Mouthparts are bluntly rounded and as broad as long, in contrast to *Psoroptes* mites which have slender, pointed mouthparts.

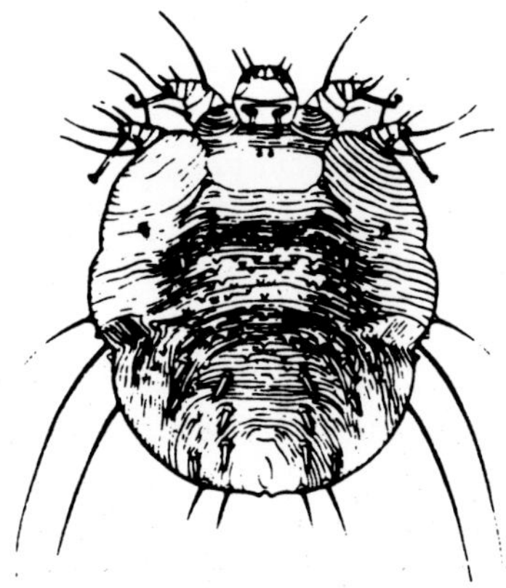

Figure 10.9. Sarcoptic mange or itch mite. (CDC)

Among livestock, this species is of greatest economic importance in large commercial swine operations. Sarcoptic mange is also the most common type of mange of dairy cattle and horses; the condition is rare in sheep and goats. Similar to psoroptic mites, the entire life cycle of *Sarcoptes* is passed on the host; however, this species burrows into the skin, forming tunnels (Fig. 10.10) in which fertilized females deposit 40-50 eggs over a 2-4-week interval. The eggs are attached in a row to the bottom of the burrow; hatching occurs within 3-5 days. The six-legged larvae may remain in the old burrow, make a new burrow or add a side tunnel to the old one; however, most apparently leave the old burrow and crawl onto the skin surface. The larval stage lasts 4-6 days. Larvae do not molt on the skin, but enter a hair follicle or make a small pocket in the skin where they molt to the nymphal stage. Nymphs also may remain in the larval burrow, enter the skin through a hair follicle or crawl onto the skin surface; those on the skin surface excavate a burrow before molting to the adult stage. Duration of the nymphal stage is approximately two days. Nymphs molt into a male or a pubescent (copulatory) female which usually remains in the burrow until fertilized. Young, fertilized females also occur on the body surface, and dissemination by direct contact to non-infested animals probably results more readily from this stage than from larvae or nymphs. After 2-4 days, the fertilized female matures to an egg-laying female which, within an hour, burrows into the skin and begins depositing eggs. Adult males wander over the skin in search of females. Adults may live up to two months. Sarcoptic mites are very susceptible to desiccation and cannot survive long periods off the host.

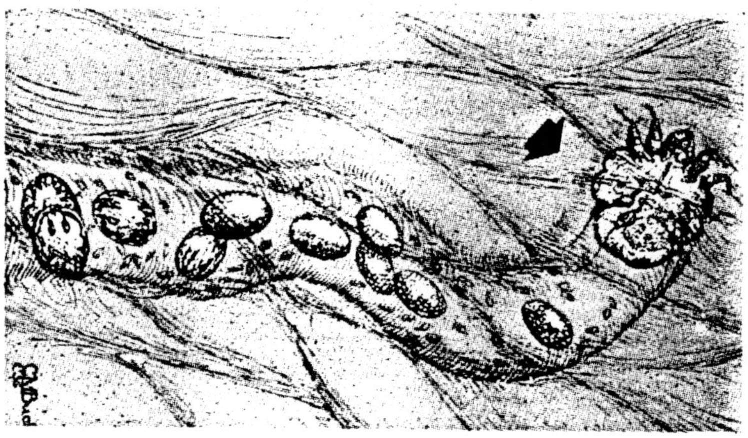

Figure 10.10. Sarcoptic mange mite burrowing under skin and laying eggs. (CCEX)

The highest incidence of sarcoptic mange normally occurs in late fall and early spring. Burrowing and feeding activity results in severe skin inflammation, accompanied by an serous exudate which hardens into a crust or scab. Lesions may appear on any part of the body, but usually begin on the head around the eyes, nose and ears or around the neck in swine. In cattle, infested areas frequently initiate above the scrotum or udder and on the inner surface of the thighs. If the disease is not controlled, it may spread over the entire body. The skin thickens and wrinkles and affected areas lose hair. Secondary bacterial infection is a frequent complication. Severe, untreated cases lead to progressive emaciation and death.

Chorioptic Mites, *Chorioptes*

Chorioptic or symbiotic mange mites (Fig. 10.11) resemble psoroptic mites but are slightly smaller in size. The body is oval and slightly longer than wide. Mouthparts are rather short and blunt in contrast to the comparatively long, sharp mouthparts of psoroptic mites. Livestock hosts include cattle, horses, sheep and goats; dissemination is by direct host contact. This species, similar to *Psoroptes*, does not burrow beneath the skin and the entire life cycle is passed on the host. *Chorioptes* mites apparently feed on skin debris rather than piercing the skin of the host as do *Psoroptes* and *Sarcoptes* mites.

Mites pass through an egg, larval, two nymphal, two female and one male stage. Egg-laying females usually deposit only one egg per day. Females that deposit eggs may live up to three weeks, whereas those not depositing eggs and males may survive 7-8 weeks. During the egg-laying process, females secrete a fluid which solidifies to fasten the egg to the host. Eggs apparently are deposited directly onto the skin. The presence of eggs from one female seemingly attracts other females to oviposit in the same locality. Under natural conditions on the host's skin, eggs require 5-6 days to hatch. The larval and two nymphal stages require 3-5 days each. Copulatory females may require up to eight days for development with a preoviposition period of about two days. The complete life cycle requires 21-26 days under laboratory conditions.

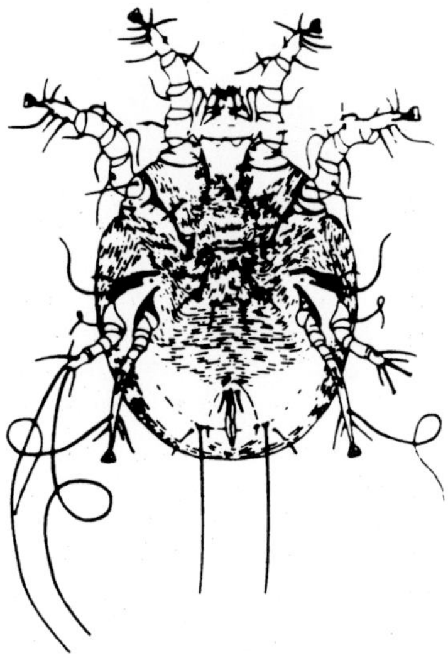

Figure 10.11. Chorioptic mange mite. (PCEX)

Chorioptic mange, like both psoroptic scab and sarcoptic mange, tends to occur during the winter months; however, the disease is milder and typically occurs on the lower hind legs and around the tail. Lesions are manifested in localized areas by small, irregular nodules that exude serum. In advanced cases, scabs may become thick and heavy and the skin becomes thickened and wrinkled as in psoroptic and sarcoptic mange. Infestations tend to disappear in the spring.

Demodectic or Follicular Mange Mites, *Demodex*

Demodectic mange mites (Fig. 10.12) infest hair follicles and glands of the skin of their host. The bodies of adult mites are elongated and appear separated into three divisions; the elongated abdomen has transverse lines on top and bottom. Legs are three-segmented, short and stubby and terminate in a pair of claws.

Figure 10.12. Demodectic mange mite (CDC)

The life history of demodectic mites is not well-known. Larvae have three pairs of legs and there appears to be three nymphal stages and the adult. Follicle mites seem fairly resistant to environmental conditions and can survive several days off the host in moist surroundings. Lesions typically appear as small, firm nodules most often around the head and neck, shoulders and genital areas; these occasionally rupture, releasing a thick, "cheesy" exudate. Infestations are apparently more common in cattle, especially dairy breeds, than other livestock species. It is generally unnecessary to treat infestations of these mites on livestock since economic loss is minimal.

MITES OF DOGS AND CATS

Demodectic and Sarcoptic Mange

Mange of dogs (Fig. 10.13) is the result of infestation by either sarcoptic or demodectic mites. The life cycles of both species are passed on the host and are similar to those described under mites attacking livestock. Both species occur beneath the skin surface, *Sarcoptes* in burrows and *Demodex* in hair follicles. Sarcoptic mange lesions may occur on any part of the body but usually appear initially on the head as small, reddish blisters; scratching by the host causes the area to spread rapidly. Burrowing of the female mite results in serous exudates which form crusts or scabs. The skin may thicken and become wrinkled, and the hair may fall out. Itching is intense; scratching may give rise to secondary bacterial infections producing an unpleasant odor. The first indication of demodectic, follicular or red mange in dogs is the appearance of lesions chiefly about the muzzle and eyes and on the front feet. Many dogs apparently harbor the mite with little or no symptomology; clinical disease appears more prevalent in older dogs and those with long hair. Mange reaction may be due to a lowering of the immune response. A bacterium, *Staphylococcus pyogenes*, is frequently associated with active lesions and may be the actual cause of the disease. Itching becomes pronounced as the lesions spread. After the invasion by bacteria, the infection becomes pustular and the skin becomes thickened; poisons formed by the bacteria affect the general health of the animal and a disagreeable odor is emitted. The condition may persist for several years and usually results in death if untreated. Differential diagnosis of mange type, made on the basis of microscopic identification of the mite from skin scrappings, is important because follicular mange is more difficult to treat.

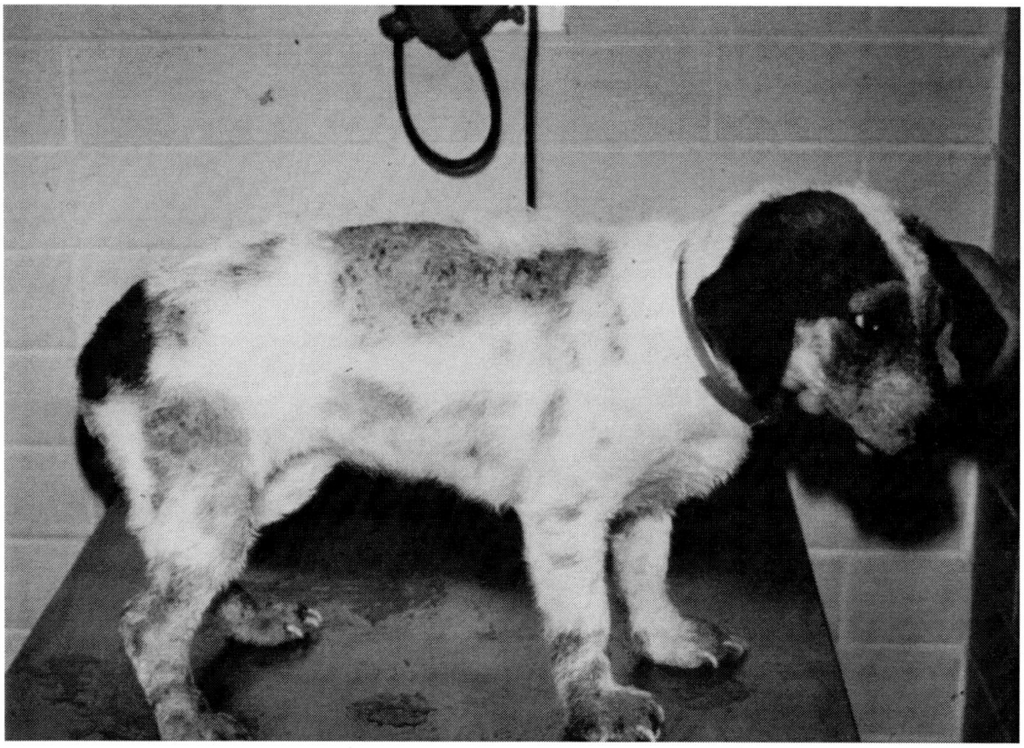

Figure 10.13. Demodectic or red mange of dog.

Notoedric Mange Mite

Notoedres cati (Fig. 10.14) is worldwide in distribution; the principal host is domestic cats. This species is similar in appearance and biology to *Sarcoptes*. The life cycle, from egg to adult, takes approximately 17 days; maturation and deposition of the first egg requires an additional 4-5 days so that the generation time is about three weeks. Ovigerous females deposit about 60 eggs at a rate of 3-4 eggs per day throughout a lifetime of 2-3 weeks. Burrowing activity of the female is responsible for the typical scab-like lesions which occur primarily about the ears, head and neck. Notoedric mange is highly contagious among cats and can result in death in 4-6 months. Treatment consists of the application of an approved acaricidal formulation.

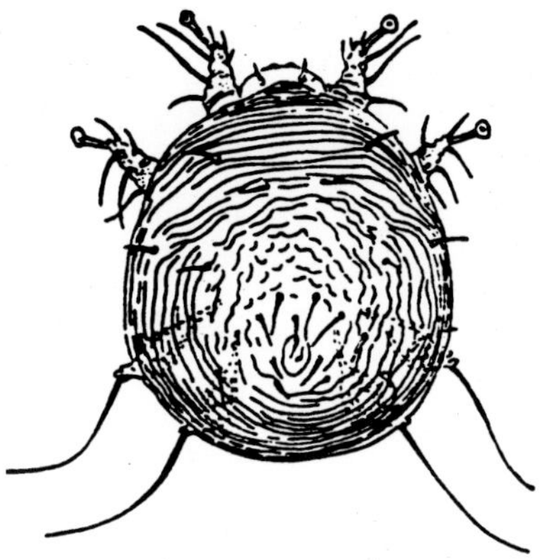

Figure 10.14. Notoedric mange mite. (USDA)

Otodectic Ear Mite

Otodectes cynotis (Fig. 10. 15) generally occurs deep in the auditory canal, near the eardrum, of dogs and cats. The life history is generally similar to that of *Psoroptes*, except that the entire life cycle is passed in the ear canal resulting in the formation of scab-like lesions in this area; infestations are usually bilateral. Dissemination is believed to occur via direct contact with an infested host. Considerable itching and scratching on the part of the host is common; convulsions may occur in heavily infested individuals. Control consists of treatment with an approved acaricidal solution following cleaning of the ear canal to remove as much as the hardened exudate as possible.

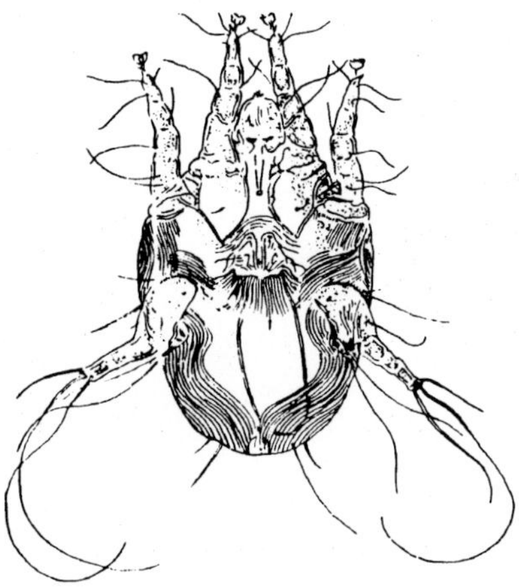

Figure 10.15. Otodectic ear mite. (USDA)

CHAPTER 11

CONTROL

The first principle of insect control is the correct identification of the pest species. Identification provides access to published information concerning the biology of the species necessary to formulate an effective control program. Once this has been established, various methods of control, alone or in combination, may be considered. The decision to initiate any control program should be based upon an assessment of the magnitude of each problem; the guiding principle in such situations is that control procedures are justified only when the expected loss without treatment exceeds the cost of the treatment itself.

INTEGRATED CONTROL

Although many insect control programs have relied almost exclusively on the use of chemicals in the past, the most successful programs have depended ultimately upon a combination of methods, termed integrated control, in which the proper and judicious use of insecticides supplements, rather than supplants, other methods of control.

Physical or mechanical control, exemplified by environmental sanitation, is of primary importance in premise fly control programs. Similarly, only a small amount of insecticidal application is necessary if good water management practices are carried out for mosquito control. Tick populations may be reduced by elimination of brush by cutting or burning.

Cultural control includes such practices as pasture vacation or pasture rotation for the control of certain species of ticks. Animal husbandry practices, such as castration and dehorning, should be performed so that wounds will have minimum exposure to flies during their season of abundance to prevent screwworm and other blow fly infestations.

Regulatory control involves the invocation of quarantine procedures to prevent the spread of pest species (e.g., *Boophilus* ticks and *Psoroptes* scab mites) in eradication programs.

The phenomenon by which certain breeds of animals are more resistant to attack by various arthropod pests is termed host resistance. Brahman cattle are generally less susceptible to attack by ticks and horn flies than are most European breeds.

The classic example of the sterile male technique is the program which eradicated the screwworm, *Cochliomyia hominivorax*, from the United States. Millions of laboratory-reared, irradiated screwworms were liberated into the field where the males mated with normal females. Since the female screwworm fly mates only once, eggs never hatched from females which had mated with sterile males. If adequate numbers of vigorous and competitive sterile males are introduced into a natural population to mate with normal females, that population will soon cease to exist. Sterile males should outnumber the fertile males in the target population so that the chances of a sterile male mating with a normal female greatly exceeds that of a fertile male. The application of the sterile male technique to control other species of insects is being investigated at the present time.

Biological control involves the use of other organisms (e.g., pathogens, parasites, predators and competitors) to reduce populations of arthropod pests. The bacterium, *Bacillus thuringiensis*, for example, may have potential use in controlling species of filth flies as may certain species of hymenopterous pupal parasites. Dung beetles bury manure that would otherwise serve as breeding media for horn flies and face flies. Mosquito fish, such as the top-feeding minnow, *Gambusia*, are effective predators of mosquito larvae.

Great progress also has been made in research on non-toxic chemicals, specifically the insect growth regulators (IGR's) such as the ecdysones and juvenile hormones. Insects have a tough outer cuticle or exoskeleton which protects the animal and serves to support and contain the internal organs. Since this exoskeleton is incapable of growth, the insect periodically replaces this protective outer skeleton with a larger one as it develops from the egg through the immature stages to the adult. This growth process is known as "molting" or "ecdysis". Two hormones are involved in the control of immature insect molting: (1) the molting hormones, or ecdysones, and (2) the juvenile hormones. Ecdysones are necessary for molting to occur, playing a key role in resorption of the old cuticle and deposition and hardening of the new one. The juvenile hormones must be present at each molt during the immature stages to prevent the insect from maturing; conversely, they must be relatively absent for the immature insect to molt to the adult stage. Ecdysones and juvenile hormones have been studied extensively and may ultimately play a role in controlling some species of insects.

Chemical control involves the use of toxic chemicals, termed pesticides, to reduce populations of economically important species of arthropods. A perfect pesticide, if such were possible, would be (1) highly toxic to the pest species; (2) harmless to man and to beneficial animals, insects and plants; (3) attractive to the pest and not unpleasant to humans; (4) inexpensive, easy to produce and readily available; (5) chemically stable for residual applications; (6) unstable for use in aerial applications that will not grossly contaminate the environment, killing pests rapidly, and breaking down subsequently into harmless compounds; (7) nonflammable; (8) noncorrosive; (9) nonstaining; and (10) easily prepared into any desired formulation. Several pesticides have some, but not all, of these desirable attributes in varying degrees.

CLASSIFICATION OF INSECTICIDES

Insecticides may be classified in a number of ways. One such way is according to the manner in which they act to kill pests. Stomach poisons generally consist of poison baits, frequently mixed with an attractant, that must be ingested to cause death. Contact insecticides penetrate the cuticle of arthropods. They include residual sprays applied to walls and ceilings of buildings to kill insects that rest on treated surfaces, aerosols and space sprays that adhere to and kill flying insects, and larvicides that penetrate the tracheae and body wall to kill insect larvae. Fumigants consist of volatile chemicals whose vapors enter the insect body through the spiracles. Desiccants are sorptive dusts which abrade the cuticle or absorb its waxy, protective outer coating causing the insect to lose body fluids and die as a result of dehydration.

Another system of classification is based on the insect stage upon which the insecticide acts. Adulticides kill adult insects; larvicides kill the immature stages of insects, whether larvae or nymphs; and ovicides kill insect eggs. Many other terms are used, such as acaricides for killing ticks and mites, and pediculicides for killing lice.

A third system of insecticide classification is based upon the chemistry of the compounds. Inorganic chemicals, such as calcium arsenate and lime-sulfur, contain some of the oldest insecticides used. Botanicals, such as pyrethrum, are extracts of naturally occurring plant products. These materials, in use before 1900, are still widely used because they pose little problem with toxic residues. Chlorinated hydrocarbons, such as methoxychlor and lindane, were the most widely used group of insecticides from the 1940's through the 1960's. However, problems with resistance and environmental contamination arose and use of this group of toxicants is now greatly restricted. The organophosphates, such as malathion and diazinon, gradually replaced the chlorinated hydrocarbons because they controlled resistant insects, were biodegradable, and did not contaminate the environment. Most, however, are relatively toxic to mammals. The carbamates, such as carbaryl, are a class of contact insecticides similar in mode of action to that of the organophosphates. Synthetic pyrethroids, such as permethrin, fenvalerate and flucythrinate, are synthetic derivatives of the naturally occurring pyrethrins. This relatively new class of contact insecticides is both biodegradable and relatively nontoxic to mammals. Insect specific insecticides include the juvenile hormone analogs, such as methoprene, and the ecdysones, such as diflubenzuron and the avermectins. Ecdysones and juvenile hormones are biodegradable and do not persist in the environment. Moreover, due to differences in the physiology between insects and mammals, these compounds apparently pose minimum environmental impact.

INSECTICIDE FORMULATIONS

Commercially available pesticides consist of a mixture, or formulation, of the toxicant, termed the active ingredient or technical material, with one or more inert ingredients such as a carrier, diluent, emulsifier or wetting agent that is generally not pesticidally active. Common formulations include dusts, wettable powders, emulsifiable concentrates and oil solutions (Fig. 11.1). Other types of formulations are basically developed from these standard formulations. Some insecticides may be adapted to more than one type of formulations while others, by reason of their chemical or physical properties, may be limited to only one. Only those formulations manufactured specifically for use on animals should be used on livestock and poultry since inert ingredients safe for plants may burn skin or cause blindness.

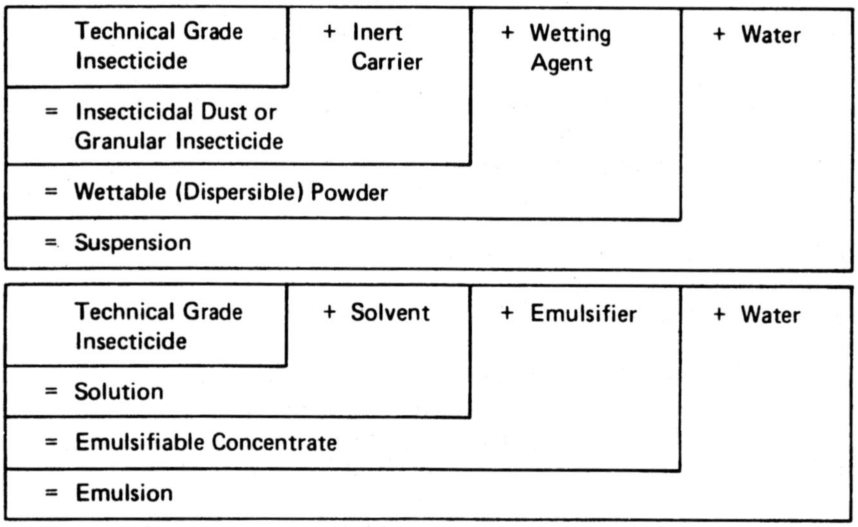

Figure 11.1. Types of insecticide formulations. (CDC)

Dusts are prepared by applying the active ingredient, dissolved in a volatile solvent, to a carrier (attapulgite. bentonite, calcite, diatomite, gypsum, hydrated lime, kaolin clay, pyrophyllite or talc); the solvent is subsequently evaporated leaving the toxicant adsorbed to the carrier material. Most insecticidal dusts are purchased ready-to-use but may be diluted for special purposes. Dusts must be of extremely fine particle size and of low density in order that they may be airborne for a considerable distance when applied.

A wettable (or water-dispersible) powder consists of the technical grade insecticide, an inert carrier and a wetting agent (usually a synthetic detergent) which facilitates mixing with water to form a suspension. The insecticide is coated (adsorbed) onto a fine inert dust, such as talc or pyrophyllite, which has thin, plate-like particles that remain suspended in water for a reasonable period of time. Many wettable powders contain an anticaking agent, which prevents lumping while in storage, and a dispersant, which helps keep the particles from settling out too quickly. During use, suspensions should be agitated continuously or frequently to prevent settling of the solid particles. Wettable powders have advantages over other concentrates. They do not require the addition of solvents which can result in injury to plants and may irritate the skin of the host. Another great advantage is the tendency of the insecticide to be deposited on the porous surface of the structure being treated. When porous materials such as concrete, plaster, adobe or unpainted wood are sprayed with a suspension, the water penetrates, leaving the carrier and the maximum amount of the active ingredient remaining on the surface as a residue. By contrast, solutions or emulsions readily penetrate porous materials so that less of the technical material remains on the surface.

An insecticide solution results when the active ingredient is dissolved in a solvent such as kerosene, diesel oil or xylene. Solutions are generally available as ready-to-use formulations or as solution concentrates. Solution concentrates contain a high percentage of technical material and must ordinarily be diluted with oil, water or other suitable solvent or diluent prior to use. Some concentrates, however, are used without dilution in ultra-low-volume (ULV) applications. Oil solutions are used extensively in fog applications.

Emulsifiable concentrates consist of the active ingredient, a solvent and an emulsifying agent, usually a synthetic detergent, which forms an emulsion when mixed with water. Emulsifiable concentrates are preferable to wettable powders for use in small spraying operations since less agitation is required to maintain a uniform mixture. Emulsions are widely used for the residual treatment of solid surfaces; insects that rest on these treated surface are killed by the residue of the active ingredient.

PREPARING INSECTICIDES FOR USE

Dusts are usually purchased ready-to-use but may be diluted for special purposes with one of the inert carriers. Wettable powders and emulsifiable concentrates, on the other hand, usually require dilution with water to a recommended finished preparation as follows:

For a Wettable Powder (WP) or Soluble Powder (SP):

$$\text{Amount of (WP) or (SP) Concentrate Required*} = \frac{\text{\% Active Ingredient in the Finished Preparation} \times \text{Number of Gallons of the Finished Preparation} \times 8.345 \text{ lb/gal}}{\text{\% Active Ingredients in the Concentrate}}$$

*The unit of weight for the answer will always be pounds (lb).

For an Emulsifiable Concentrate (EC):

$$\text{Amount of (EC) Concentrate Required*} = \frac{\text{\% Active Ingredient in the Finished Preparation} \times \text{Quantity of the Finished Preparation}}{\text{\% Active Ingredients in the Concentrate}}$$

*The unit of volume for the answer will always be the same unit used for the "quantity of the finished preparation".

INSECTICIDE APPLICATION METHODS

Historically, livestock treatment methods included dipping animals in chemically-charged vats; single animal or herd treatment by high pressure sprayers; hand application of solutions, smears or dusts to affected body areas; or the use of self-applicatory devices such as backrubbers or "oilers". More recently, the results of research on new formulations of certain pesticides and insect growth regulator (IGR) compounds have provided easy and laborsaving, but not necessarily as economical, application techniques.

Dusts may be manually applied to livestock and poultry by hand shakers or hand-pump dusters; however, self-treatment dust bags (Fig. 11.2C) are more common. For parasites within the ears, a rubber-tipped, spring-operated oil can may be used.

Sprays (Fig. 11.2B) are used primarily for the control of lice, ticks and mites on range cattle; animals must be wet to the skin when applying systemic insecticides for cattle grub control. Sprayers include a wide range of design from portable, hand operated units to large, truck mounted, power-driven equipment. Pressure regulating devises may be required to maintain pressures of 250-350 psi. Sprayers can be moved easily from pasture to pasture; however, transporting water may be a problem because up to one gallon of spray per animal is often required for effective parasite control.

Because animals are wet thoroughly in a vat, dipping (Fig. 11.2A) provides an extremely uniform method of coverage. Dip vats may be permanent concrete, metal or wood frame structures, or they may be portable units mounted on trailers for towing between ranches. Vats can be filled early in the season and used several times during the year. Although initial cost of filling and charging the vat is high, cost per animal may be relatively low since many animals can be treated repeatedly during the season with little additional expense. When filling or recharging a vat, only formulations specially prepared for dipping animals should be used.

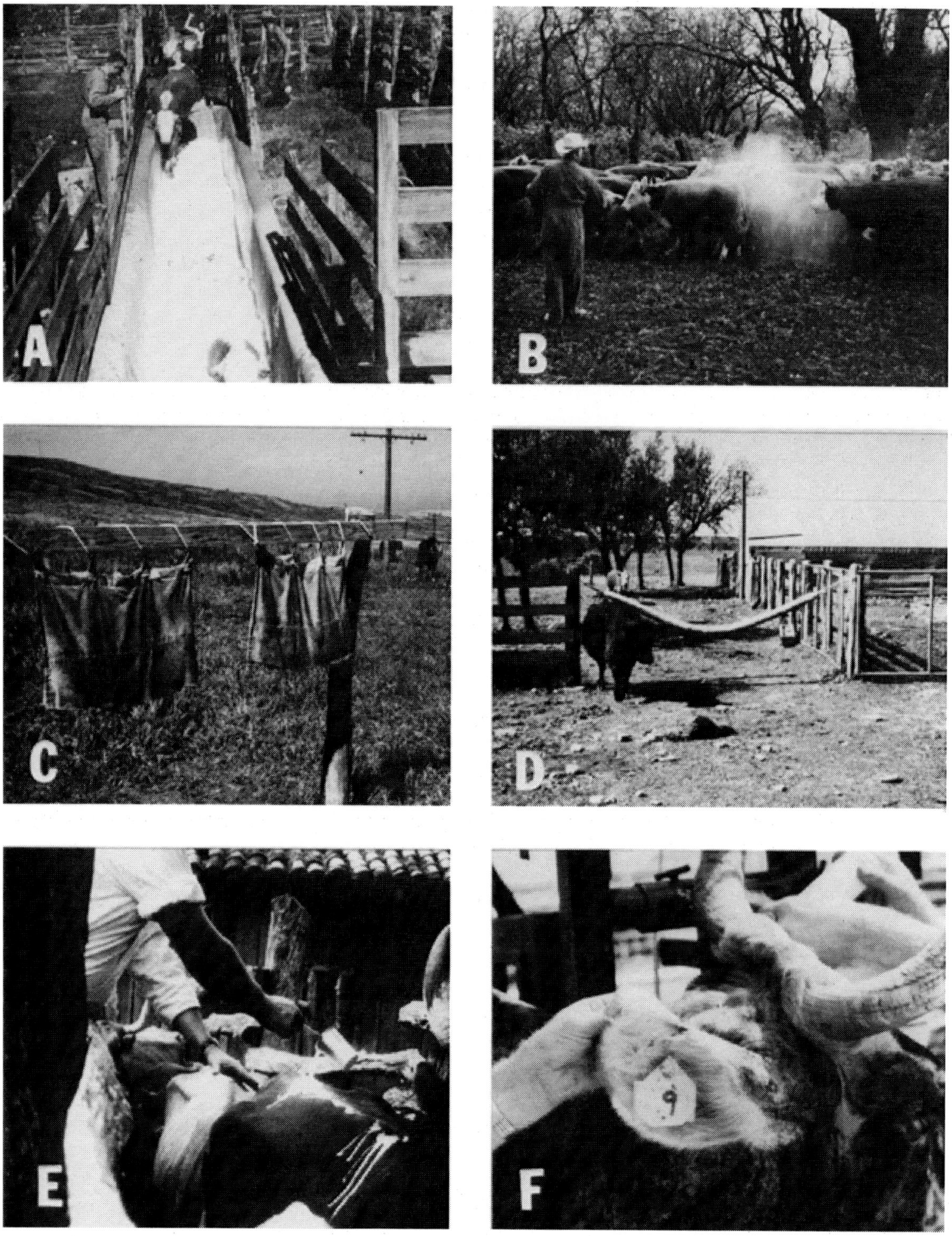

Figure 11.2. Insecticide application methods: A. Dip vat, B. Spray, C. Dust bag, D. Backrubber, E. Pour-on, F. Ear tag. (USDA)

Self-applicatory devices (Fig. 11.2C, D) consist of backrubbers, oilers and dust bags designed to aid in the control of horn flies, ticks and lice on range cattle and dairy-pastured cows. Such devices must be employed in "forced-use" situations for maximum effectiveness. Backrubbers vary in design from simple ropes or cables wrapped with insecticide-treated burlap sacking to portable, metal constructed units with vertically positioned supply tanks. Home-made dust bags may be constructed from burlap sacks filled with insecticide dust and hung in trees or other suitable locations; commercial dust bag design provides a stronger and more durable bag as well as plastic inner liner to prevent clumping of the dust due to moisture.

Another innovative method, originally designed for ear tick control, is the use of insecticide-impregnated plastic ear tags (Fig. 11.2F) for cattle. The active ingredient is incorporated into the plastic during the manufacturing process and is gradually released through contact with the animal's hair coat following application. Ear tags are also useful in controlling horn flies on cattle in areas where resistance has not developed. Flea collars for dogs and cats are similarly formulated.

High concentrate, low-volume backline pour-on and spot-on formulations (Fig. 11.2E) were initially developed for systemic control of cattle grubs; however, such treatments also are effective against lice and horn flies, although control of these latter pests is rather short lived. These materials are applied by means of a small metal dipper or by an automatic refillable syringe or plastic bottle. Recommended products may be purchased in ready-to-use or water-miscible formulations and applied to animals at rates from 0.5 to 1 ounce per 100 pounds of body weight up to a maximum of 4-8 ounces per animal, depending upon the material selected.

Control of certain fly species, such as horn flies and face flies, whose larval stages develop in animal manure may be obtained when preparations of nonsystemic insecticides and insect growth regulating compounds (IGR's) are consumed by the animal as feed or mineral additives. The active ingredient of these materials passes through the intestine unaltered and is voided in the feces. Optimal daily intake is difficult to achieve with cattle exposed to free-choice feeding of supplement rations on rangeland; greater success is obtained with animals under confinement-type operations (drylot dairies and beef feedlots) where controlled daily feeding is practiced. Recently developed bolus formulations of these compounds are also useful as larvicides. Systemic materials administered in this manner are effective for grub and louse control.

Baits are primarily used as an aid in the control of house flies around feedlots, dairies, poultry houses, livestock barns and similar places where flies tend to congregate. Baits consist of attractive substances (e.g., dry sugar or syrup) to which a small amount of insecticide is added to kill flies feeding on the mixture.

Insecticides formulated as pastes usually are to be applied orally. Paste treatments are formulated in a ready-to-use syringe. Label instructions should be followed closely.

The new avermectin materials are marketed as subcutaneous injections. At present, they are recommended for the control of grubs, sucking lice and scabies mites as well as certain species of parasitic helminths.

INSECTICIDE RESISTANCE

Most populations of insects contain individuals that vary in their susceptibility to insecticides; individuals that are less susceptible are referred to as more tolerant of the chemicals used. Continued insecticidal pressure to a population will eventually destroy the more susceptible individuals, permitting the more tolerant individuals to survive and reproduce new generations of increasingly less susceptible insects. Thus, the species becomes increasingly more difficult to control because of genetic factors transmissible to subsequent generations.

Resistance of insects to insecticides is defined as the ability to withstand a toxicant which was previously lethal to earlier populations. In general, two types of resistance occur in insects. Physiological resistance is the ability to withstand, through physiological processes, a toxicant due to differences in (1) the permeability of the insect exoskeleton to insecticides, (2) the detoxification of insecticides into less harmful compounds, (3) the storage of insecticides harmlessly in body tissues such as fat, or (4) the excretion of insecticides. Behavioristic resistance is the ability to avoid lethal contact with the insecticide through protective habits or behavior, such as mosquitoes resting outdoors rather than on treated surfaces, or flies avoiding baits containing organophosphates.

Resistance to insecticides has been reported in many insects. With some species, initial resistance to one insecticide soon leads to resistance to other related chemicals as well; this phenomenon is known as cross-resistance. Cross resistance to some compounds occurs to such an extent among populations of certain species of arthropods that these chemicals are no longer of value in controlling these pests throughout many areas of the United States.

PESTICIDE REGULATIONS

Following the publication of Rachel Carson's famous book Silent Spring in 1962, increasing restrictions were placed on the use of insecticides by federal, state and local agencies. Continuing pressure was exerted to limit the use of pesticides and to combine all activities of the Federal Government concerned with insecticides and their effects on the environment into a single agency. As a result, Congress passed legislation creating the Environmental Protection Agency in 1970. The EPA includes organizations concerned with pesticides, or their effect on the environment, which were previously parts of the Departments of Agricuiture, Interior, and Health, Education and Welfare; the Council on Environmental Quality; and the Atomic Energy Commission.

The Federal Insecticide, Fungicide and Rodenticide Act (FIFRA), first passed in 1947, was significantly amended by the Federal Environmental Pesticide Control Act (FEPCA) of 1972. This new law extended federal registation and regulation to all pesticides, including those distributed or used within a single state. Its major features (1) require the proper application of pesticides to insure greater protection of man and the environment, (2) prohibit the use of any registered pesticide in a manner inconsistent with labeling instructions, (3) specify that insecticides must be classified for "general use" and "restricted use", or both, and (4) mandate that federal standards for certification of individuals who apply restricted pesticides be set forth by the EPA.

General use pesticides are those, which when applied in accordance with their directions for use, warning and cautions, or for one or more of such uses for which they are registered, or in accordance with widespread and commonly recognized practice, will not generally cause unreasonable adverse effects to the environment or man. This group includes those insecticides with low human toxicity such as pyrethrum, rotenone and methoxychlor, or diluted ready-to-use formulations of low toxicity materials such as malathion which can be applied by the general public.

Restricted use pesticides are those, which when applied in accordance with their directions for use, warning and cautions, or for one or more of such uses for which they are registered, or in accordance with widespread and commonly recognized practice, may generally cause without additional regulatory restrictions unreasonable adverse effects to the environment including injury to the applicator. These pesticides can be applied only by, or under the direction of, a certified applicator. Standards for federal certification of pesticide applicators are established by EPA. An individual must establish that he is competent in the use or handling of pesticides to be certified. This group of pesticides includes highly toxic materials such as parathion; concentrates of moderately toxic materials such as chlordane and diazinon, which might pose a hazard to operators diluting the toxicant to field stength formulations; and many insecticides which may contaminate the environment, such as mosquito larvicides which might affect aquatic organisms and therefore should be applied only by trained personnel.

The decision to apply any pesticide is the responsibility of the agency or the individual concerned. Anyone unfamiliar or inexperienced with a specific procedure should seek the advice of personnel experienced in its use or obtain the necessary training from an authorized source. AT THE PRESENT TIME, THE BEST ADVICE TO ALL PESTICIDE APPLICATORS IS TO CONSULT AN APPROPRIATE AUTHORITY OR STATE REGULATORY AGENCY BEFORE APPLYING AN INSECTICIDE WHOSE USE MIGHT BE IN QUESTION.

PESTICIDE LABELS AND LABELING

A pesticide label (Fig. 11.3) is the end product of discovery, research, development, and the federal and state registration process. Federal law defines the label as the written, printed or graphic matter on, or attached to, the pesticide or any of its containers or wrappers. The importance of READING THE LABEL cannot be overly stressed. The information that appears there is included for your information and your protection. If it is read, understood and all directions followed, the likelihood of misusing the material or of having an accident is remote. The label is a tool; learn to use it.

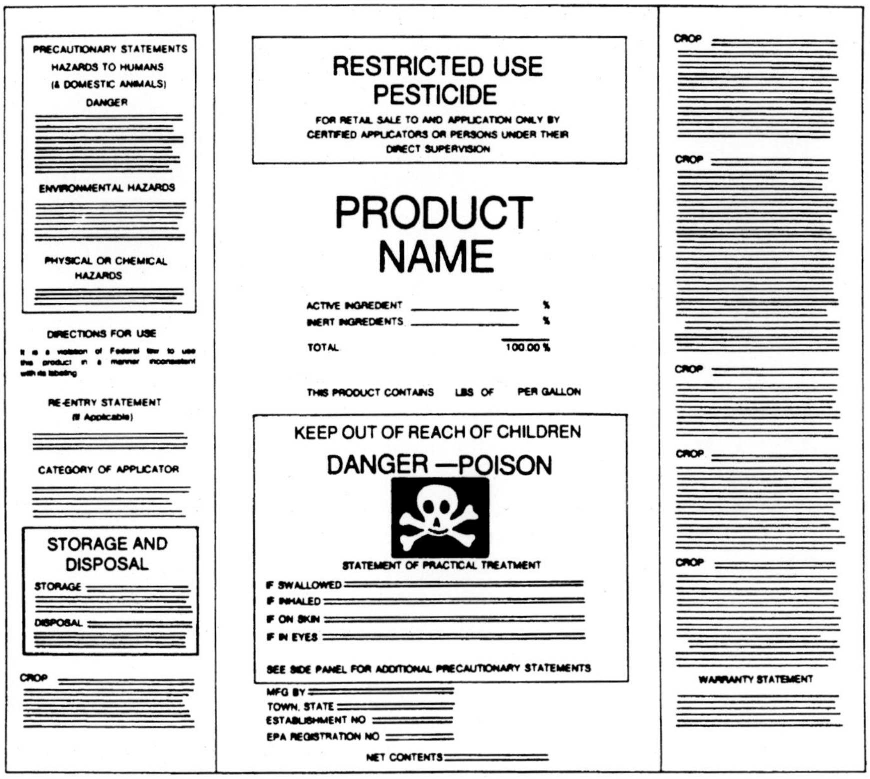

Figure 11.3. Sample pesticide label. (TAEX)

The printed material on a pesticide label has all of the necessary information and instructions for the effective and safe use of a pesticide. It must, by law, include the following information: (1) brand name; (2) type of formulation; (3) active ingredients; (4) inactive ingredients; (5) net contents in U.S. units; (6) EPA registration and establishment numbers; (7) name and address of manufacturer ot registrant; (8) general-use or restrictive-use classification; (9) directions for use, including pests to be controlled; animals or sites to be treated; dosage, time, and method of application; limitations relating to the time between the last application and slaughtor to the marketing of milk; and directions for storage and disposal; (10) precautionary statements to include physical or chemical hazards associated with the product, environmental hazards, and hazards to humans; (11) statement of practical treatment; (12) signal words and symbols. The word "Danger" identifies a highly toxic product; highly toxic products also must carry the "skull-and-crossbones" and a section informing physicians of the proper treatment for poisoning. The signal word "Warning" is required on moderately toxic products, while "Caution" is required on all products of low toxicity. All products must also bear the statement, "Keep out of the reach of children".

PRECAUTIONS FOR THE SAFE USE OF PESTICIDES

Most cases of overt poisoning by pesticides occur in workers applying these agents. Therefore, the safe handling of insecticides is a necessity for persons applying insecticidal dusts, whether the materials used are highly toxic or only slightly toxic to humans. Respiratory exposure occurs most frequently among persons that handle dusts which are not readily absorbed through the skin. Dermal exposure becomes much more hazardous when applying sprays. Solvents, such as xylene, used with many sprays destroy the protective oily barrier of the skin and allow more intimate contact with the chemical toxicant. Poisoning by ingestion occurs most frequently among children under five years of age when pesticides are kept in accessible places.

1. Read the container label and understand the directions for the preparation and application of the insecticide, then follow all directions and precautions precisely.

2. Store insecticides in containers with the original labels. Keep insecticides out of the reach of children and animals, preferably in a locked cabinet or box in a locked room with signs, such as "Dangerous-Chemicals - Keep Out".

3. Mix insecticides in a well-ventilated area, preferably out-of-doors, to avoid inhalation of dusts and fumes.

4. When mixing and applying flammable chemicals, avoid the fire hazard associated with tobacco smoking, defective wiring, open flames and hot plates.

5. Do not eat or smoke when working with insecticides. This practice will reduce opportunities for accidental ingestion of toxic materials.

6. Wear protective clothing and headgear and avoid prolonged exposure to the insecticide. Special care should be taken to prevent contamination of the skin when concentrates are being handled. Use impermeable gloves and aprons and wear approved respirators when necessary.

7. Avoid contamination of food and drinking water of both man and animals.

8. Avoid inhaling sprays and dusts when insecticides are being mixed or applied and avoid spilling on the skin. If any material contacts the skin, wash the affected area thoroughly with soap and water. Wash routinely after each day's application. If any particles or drops of insecticide accidentally enter the eye, flush them out immediately with large volumes of clean water.

9. If clothing becomes contaminated, remove it immediately and wash the affected part of the body thoroughly with soap and water and put on clean clothing. Wear clean clothing every day.

10. Keep insecticidal equipment clean and in good operating condition. Flush out equipment only where contamination will not affect man or wildlife.

11. Mix only as much insecticide as is needed for each application, thus reducing the problems of storing and disposing of excess material. Dispose of excess insecticide carefully so that contamination will not affect man and wildlife.

12. Dispose of empty containers safely (e.g., bury them under at least two feet of soil). In some situations, burn empty insecticide bags and stay out of the smoke. For non-returnable liquid containers, wash inside and outside thoroughly, immediately after use. Puncture large metal containers so that they cannot be reused.

13. Be familiar with first aid measures and antidotes for insecticides and the location and telephone number of a local poison control center. Follow instructions given in the statement of practical treatment given on the label.

INDEX